Inhalt
Contents

The Space That Time Forgot

Angela Bulloch

Städtische Galerie im Lenbachhaus und Kunstbau, München

Verlag der Buchhandlung Walther König, Köln

Vorwort

Schon seit geraumer Zeit hat das Lenbachhaus mit großem Interesse die Entwicklung des Werks von Angela Bulloch verfolgt. Deshalb sind wir besonders froh, mit Hilfe eines privaten Mäzens, *Z Point,* eine der wichtigsten Installationen von Angela Bulloch, als Dauerleihgabe für unsere Sammlung gewinnen zu können.
Seit langem gibt es auch den Wunsch, unserer Begeisterung für das Werk von Angela Bulloch durch eine großangelegte Einzelpräsentation im Kunstbau des Lenbachhauses Ausdruck zu verleihen. Wir sind daher außerordentlich froh und dankbar, dass dieses Vorhaben nun in enger Zusammenarbeit mit der Künstlerin umgesetzt werden konnte. Besonders stolz sind wir jedoch, dass die Ausstellung *The Space That Time Forgot* im Wesentlichen aus Arbeiten besteht, die im Kunstbau zum ersten Mal der Öffentlichkeit vorgestellt werden. Denn gerade die Herausforderung an die Künstler, ihre Neuproduktion auf die architektonischen, historischen und strukturellen Bedingungen des anspruchsvollen Kunstbaus abzustimmen, hat zu den erfolgreichsten Präsentationen von Gegenwartskunst im Lenbachhaus geführt. Im Falle von Angela Bulloch ist diese Verfahrensweise ein besonders glückliches Konzept, da mit der Kategorie „Raum" einer der wichtigsten Parameter in ihrem Werk benannt ist. Seit langem setzt sich ihr vielgestaltiges Oeuvre, bestehend aus Installationen, Soundarbeiten und Werken in zahlreichen Medien, mit den konzeptionellen, institutionskritischen wie ästhetischen Problemen der Kategorie „Raum" auseinander. *The Space That Time Forgot* ist, wie der Titel nahe legt, eine weitere Beschäftigung mit diesem Thema.
Die Publikation diskutiert das Ausstellungsprojekt im Kunstbau des Lenbachhauses, indem sie einerseits in Installationsfotografien die Ausstellung dokumentiert und andererseits in begleitenden Essays erläutert. In seinem Beitrag konstatiert Diedrich Diederichsen zunächst eine Krise des Raums, die sich sowohl auf die Vorstellung des Weltalls als auch auf das Raumgefühl des Alltags erstreckt. Diederichsen stellt die Werke von Angela Bulloch als eine Vermittlung zwischen abstraktem und konkretem Raum, zwischen einem mathematischen Erhabenen und den Sinneseindrücken, zwischen Platonischen Körpern und chaotischen Sternbildern, ja zwischen Zeichen und sinnlicher Vorstellung und so als Phänomenologie eben jener zuvor benannten Raumkrise dar.
Der zweite Essay bemüht sich um die präzise Deskription der ausgestellten Werke von Angela Bulloch im Kunstbau und versucht hinsichtlich der im Titel *The Space That Time Forgot* genannten Themen von Raum und Zeit, die wissenschafts-, philosophie- und kunstgeschichtlichen Referenzen aufzuzeigen und dem weiten Horizont der kulturhistorischen Themenvielfalt im Werk der Künstlerin gerecht zu werden.
Die Fotografien von Carsten Eisfeld, die in enger Zusammenarbeit mit Angela Bulloch entstanden, dokumentieren die in der Ausstellung gezeigten Werke. Sie heben dabei

den installativen Aspekt und den Bezug der Werke auf den Ausstellungsraum in geradezu kongenialer Weise hervor. Schließlich ist erstmals in einem Katalog zu Angela Bulloch eine umfassende Bibliographie publiziert, die aufzeigt, in welchem Umfang und in welcher Qualität die Arbeiten von Angela Bulloch diskutiert werden – zugleich ist diese Bibliographie ein wichtiges Instrument zur weiteren Diskussion und Forschung.

Allen am Katalog Beteiligten gilt unser großer Dank. Zu nennen sind vor allem Angela Bulloch und Carsten Eisfeld, die das Konzept und das Design für diese Publikation entwickelt haben. Ganz herzlich gedankt sei auch Diedrich Diederichsen für seinen äußerst anregenden Beitrag, der uns gleichermaßen inspiriert wie unsere Perspektive auf die neusten Arbeiten von Angela Bulloch enorm erweitert hat. Der Verlagsbuchhandlung Walther König gilt unser Dank, dass dieses Buch mit großer Begeisterung in das Programm des Verlages aufgenommen wurde und so die langjährige fruchtbare Zusammenarbeit fortgesetzt werden konnte.

Eine solch umfangreiche Ausstellung kann nur in der Zusammenarbeit und mit Hilfe Vieler zum Erfolg geführt werden. Vor allem die erstmalige Zusammenarbeit zwischen Angela Bulloch und dem Musiker Michael Iber führte zum Gelingen der eigens für die Ausstellung entwickelten Soundarbeit *Z Phrase,* dafür sei ihm herzlich gedankt. Für ihre ideelle und organisatorische Unterstützung möchten wir Esther Schipper in Berlin ganz herzlich danken. Unser aufrichtiger Dank gilt allen Mitarbeiterinnen und Mitarbeitern des Lenbachhauses, die viel Engagement und Begeisterung in dieses Projekt eingebracht haben. Den Leihgebern Esther Schipper in Berlin, Galerie Micheline Szwajcer in Antwerpen, Galerie Eva Presenhuber in Zürich, Galería Helga de Alvear in Madrid, Cristina Guerra Contemporary Art in Lissabon und Collezione Nunzia e Vittorio Gaddi in Lucca sei für die großzügige Bereitstellung der Leihgaben und Unterstützung der Produktion gedankt.

Wir fühlen uns sehr geehrt, dass Angela Bulloch uns ihr Vertrauen geschenkt hat und wir die Ausstellung sowie die begleitende Publikation gemeinsam mit ihr realisieren konnten. Beides wäre ohne ihr inspirierendes Denken und intelligentes Engagement in dieser Qualität nicht möglich gewesen. So war es auch ihre humorvolle Zuversicht, welche die vielen am Projekt Beteiligten und den gewaltigen logistischen Apparat dieses Ausstellungsprojekts auf sympathische Weise zusammengehalten hat. Deshalb möchten wir Angela Bulloch hier noch einmal ganz ausdrücklich unseren tief empfundenen Dank aussprechen.

Helmut Friedel   Matthias Mühling

Foreword

The Lenbachhaus has been taking a close interest in the work of Angela Bulloch for some time now. We are therefore delighted that with the assistance of a private patron we have been able to acquire *Z Point,* one of Bulloch's most important installations, as a permanent loan.

We have also long wanted to underline our own fascination for the work of Angela Bulloch with a large solo exhibition in the Kunstbau at the Lenbachhaus. It is a great pleasure and privilege to now be able to present a show that has been designed in close collaboration with the artist. We are proud that this exhibition, "The Space That Time Forgot", consists largely of new works. The most successful exhibitions of contemporary art in the Kunstbau have always profited from artist's meeting the challenge of adapting their new works to the architectural, historical and structural conditions of this demanding venue. In the case of Angela Bulloch, this method is particularly appropriate, since the category of space is one of the most important parameters of her oeuvre. For many years her diverse installations, and her works with sound and in many other media have taken issue with the conceptual and aesthetic problems associated with the category of "space", and have also looked at "space" in terms of institutional critique. "The Space That Time Forgot" is, as the title suggests, a further investigation of this subject.

This catalogue looks at Bulloch's Kunstbau exhibition project in two ways: by documenting the show through installation photos, and also in accompanying essays. In his text, Diedrich Diederichsen first notes that there is a current crisis of space, both in terms of our conception of the universe and our experience of the space of our everyday lives. Diederichsen presents Angela Bulloch's works as a form of communication between abstract and concrete space, between a mathematical sublime and our sensory perceptions, between Platonic bodies and chaotic star constellations—and between the sign and the experience of the senses. This all amounts to a phenomenology of the crisis in our concept of space that Diederichsen established at the outset. The second essay works on the basis of a precise description of Bulloch's works on show in the Kunstbau, and draws on the two elements of the exhibition title, space and time, identifying scholarly, philosophical and art historical references in Bulloch's work, and thus paving the way for an appreciation of the great variety of cultural and historical material that Bulloch draws on.

Carsten Eisfeld's photographs, which were created in close collaboration with Angela Bulloch, provide a documentation of the works on show. They are a marvellous record of the installational aspect of the show and the way in which the works refer to the exhibition space. Finally, this publication is the first to include a comprehensive bibliography on Angela Bulloch, showing the extent and the quality of reception of her work and providing an important resource for further discussion and research.

We would like to thank everyone who was involved in the making of this catalogue. First of all, Angela Bulloch and Carsten Eisfeld, who together created the catalogue structure and design. Warm thanks also go to Diedrich Diederichsen for his fascinating and inspiring essay, in which he has greatly broadened the perspective on Bulloch's most recent work. We would also like to thank the publisher, Verlagsbuchhandlung Walther König, for enthusiastically including this book in their list, continuing our longstanding and highly fruitful cooperation.
An exhibition as comprehensive as this can only be successful in cooperation with and involving the help of numerous individuals. For the first time the musician Michael Iber has collaborated with Angela Bulloch on the very successful development of *Z Phrase* especially for this exhibition, and we would like to thank Michael Iber very much. Many thanks also to Esther Schipper in Berlin for their support with ideas and organisation. We are also grateful to all staff at the Lenbachhaus, who worked on this project with great commitment and enthusiasm. We would like to thank all those who generously loaned works for their support of this show: Esther Schipper in Berlin, Galerie Micheline Szwajcer in Antwerp, Galerie Eva Presenhuber in Zurich, Galería Helga de Alvear in Madrid, Cristina Guerra Contemporary Art in Lisbon, and Collezione Nunzia e Vittorio Gaddi in Lucca.
We are truly honoured that Angela Bulloch has entrusted her work to us, and agreed to work with us for this exhibition and catalogue. It would not have been possible to achieve such quality without her intelligence, inspiring ideas and commitment. Frequently, it was her humorous and appealing confidence that held the many people involved in this project and its remarkable logistical apparatus together. We would like once again to express our heartfelt gratitude to Angela Bulloch.

Helmut Friedel　　Matthias Mühling

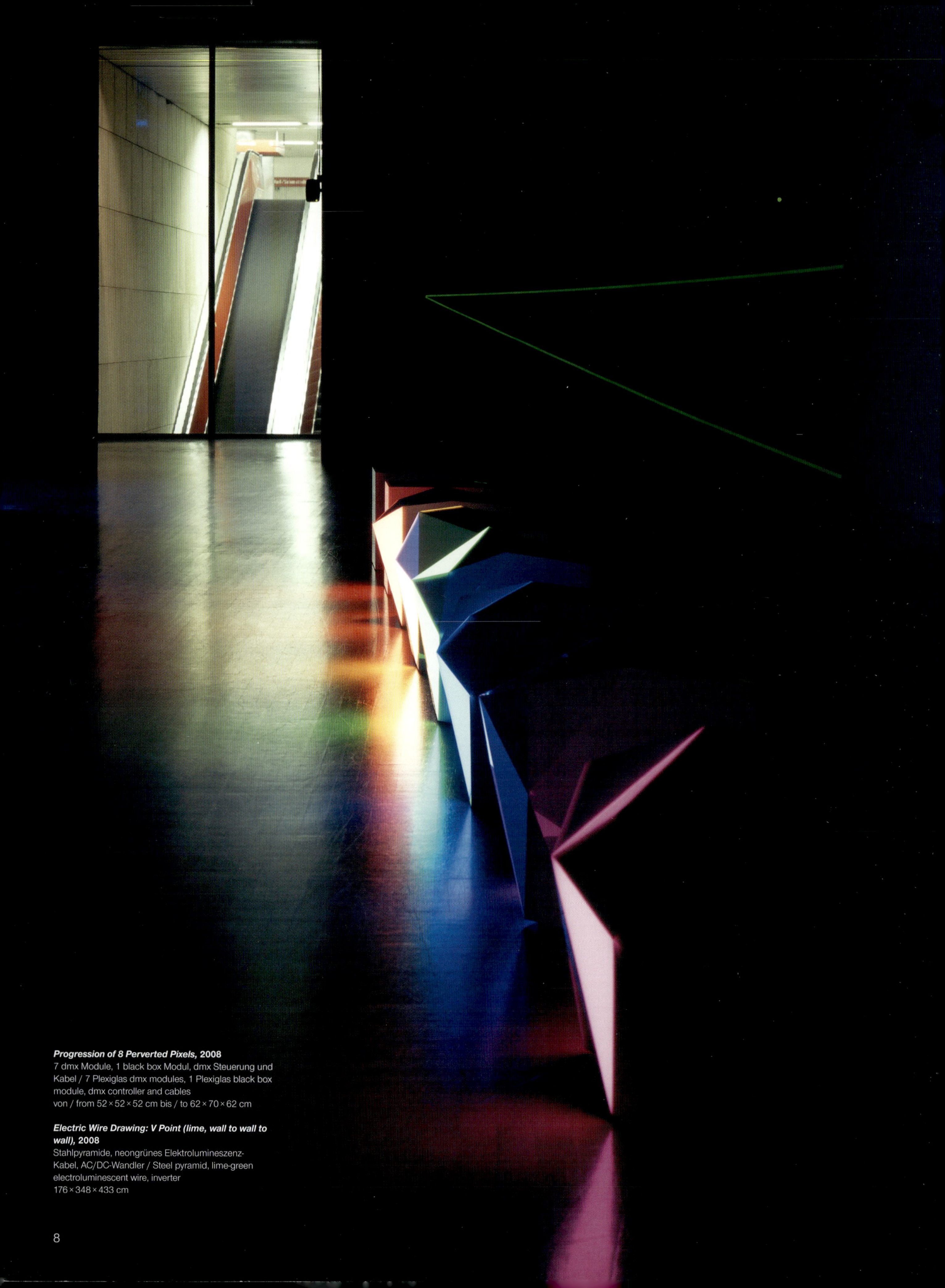

***Progression of 8 Perverted Pixels*, 2008**
7 dmx Module, 1 black box Modul, dmx Steuerung und Kabel / 7 Plexiglas dmx modules, 1 Plexiglas black box module, dmx controller and cables
von / from 52 × 52 × 52 cm bis / to 62 × 70 × 62 cm

***Electric Wire Drawing: V Point (lime, wall to wall to wall)*, 2008**
Stahlpyramide, neongrünes Elektrolumineszenz-Kabel, AC/DC-Wandler / Steel pyramid, lime-green electroluminescent wire, inverter
176 × 348 × 433 cm

Wer wohnt im schwarzen Loch?

Diedrich Diederichsen

Wir haben eine Raumkrise. War das Weltall früher unermesslich groß, aber darin einfach nur ein sehr großer Raum, sprechen seine Dimensionen heute jeder Vorstellbarkeit Hohn. Raum ist in anderen Raum eingerollt, eingekapselt, unendlich groß und unendlich klein und ortlos zugleich, sechsdimensional, neundimensional. Abzählbare Dimensionen sind aber etwas anderes als ein unermessliches All, dessen Interface Abendhimmel sich romantisch betrachten lässt. Sie geben dem Unvorstellbaren eine seltsame Konkretheit, mit der sehr viel schwerer umzugehen ist, als mit den alten Gefühlen der Erhabenheit. Diese Krise setzt sich fort im Raumgefühl des Alltags. Die Stadtplanung spricht von Angsträumen und meint damit alle tunnelartigen, unübersichtlichen und meist der Infrastruktur von Verkehr und Transport dienenden, oft unterirdischen Konstruktionen wie Parkhäuser, U-Bahn-Stationen und Unterführungen, deren Strukturen interessanterweise auch immer wieder in die Metaphorik der Astronomie eingehen: Wurmlöcher, kosmischer Verschiebebahnhof etc. Schließlich betrifft diese Krise aber auch die Kunst, und zwar auf zweierlei Weise. Zum einen kennt auch sie nicht mehr den vorgegebenen, neutralen Raum, der sich standardisiert und unmarkiert im Hintergrund hält und den Kunstobjekten den Vordergrund einräumt. Zum anderen ist aber auch die Art und Weise, wie Kunstobjekte sich auf andere Räume beziehen, fraglich geworden, wenn ihre eigene Dreidimensionalität buchstäblich nicht mehr alles umfasst, was uns an Räumlichkeit betrifft.

Raum und Zeit sind a priori, so nennt sie Kant, Voraussetzung jeglicher menschlicher Vorstellung. Dinge können entweder nacheinander geschehen oder aber neben und über einander gleichzeitig. Raum ist Voraussetzung für die Vorstellung von Simultaneität. Zeit für Sequenz, Ursache und Wirkung, Folge. Beide Begriffe führen natürlich auch als Wörter ein Alltagsleben. Sie sind von einer spezifischen Etymologie verunreinigt und von lokalen Gebrauchsweisen in spezifischen Kontexten pragmatisch geprägt. Der Raum ist im deutschsprachigen Bereich neben der Abstraktion „Dreidimensionalität“ auch für konkrete Räume zuständig – im Gegensatz zu anderen Räumen. „Zimmer“ z. B. kann im Osten Deutschlands in bestimmten Situationen durch „Raum“ ersetzt werden: etwa bei der berühmten Zwei-Raum-Wohnung, von der man in der DDR sprach. Dass man über genügend Raum verfügt, heißt in vielen deutschsprachigen Kontexten, dass man „Platz hat“, also über einen offenen Raum verfügt. Daraus folgt, dass Raum normalerweise geschlossen ist. Die Kantsche Welt, in der Raum a priori aller Vorstellungen ist, spielt in einem geschlossenen Universum. Irgendwo hinter ihm in der Zeit steht das Platonische Universum, in dem die Ordnung des Universums sich direkt in der Geometrie kleinster Körper wiederspiegelte.

Das englische Abstraktum heißt hingegen Space, es kommt auch zum Einsatz, wenn man Kant auf Englisch nachschlägt. Der dem deutschen Abstraktum Raum etymologisch

Who Lives in the Black Hole?

Diedrich Diederichsen

We are presently in the midst of a crisis of space. Formerly, the universe was immeasurably large, but it was still just a very large space; today its dimensions defy any and all attempts to conceive of them. Space is rolled up and encapsulated inside other space; it is infinitely large and infinitely small and spatially anonymous at the same time, six-dimensional, nine-dimensional. But countable dimensions are something quite different from an immeasurable universe whose interface—the evening sky—is a possible object of romantic contemplation. They lend a peculiar concreteness to the inconceivable, a brand of concreteness that is much more difficult to deal with than the old feelings of grandeur and sublimity. This crisis continues in our everyday sense of space. City planners speak of spaces of fear, by which they mean all those confusing and disorienting, tunnel-like, often underground constructions—multistory parking garages, subway stations, pedestrian underpasses—whose structures, interestingly enough, are often described with metaphors from astronomy: wormholes, cosmic switchyards, etc. Finally, this crisis also has an impact on art, in two different ways. First, art too is deprived of that preexisting, neutral space that stays in the background, standardized and unmarked, and leaves the foreground to the art objects. Second, however, the way that art objects relate to other spaces is also placed in question, since their own three-dimensionality literally no longer encompasses all the aspects of spatiality that concern us.

According to Kant, space and time are the a prioris of all human experience. Things can happen either after one another or else next to and on top of one another, at the same time. Space is the a priori for the experience of simultaneity, time the a priori for that of sequence, cause and effect, succession. Of course, both of these concepts also lead everyday lives as words. They are contaminated by specific etymologies and bear the practical stamp of local usages in specific contexts. In the German-speaking world, *Raum* refers not only to the abstract notion of three-dimensionality but also to concrete *Räume,* or spaces, in contrast to other spaces. In Eastern Germany, there are certain situations in which *Raum* may be used in place of *Zimmer,* or room, for example, in the case of the famous *Zwei-Raum-Wohnung,* or two-room apartment, that used to be spoken of in the old GDR. In many German-speaking contexts, to have enough *Raum* means to have *Platz,* or room, that is, to have an open space. This implies that *Raum* is normally closed and self-contained. The Kantian world, in which *Raum* is an a priori of all human experience, is set in a closed and self-contained universe. Somewhere behind it in time, there stands the Platonic universe, in which the order of the universe was directly reflected in the geometry of the smallest solids.

In English, by contrast, the abstract noun is *space,* which is also used in translations of Kant. The word *room,* which is etymologically close to the abstract German *Raum,* is not abstract at all; it is a concrete noun, and often an especially narrow one. The abstract

nahe stehende Room ist kein Abstraktum, sondern ein oft besonders enges Konkretum. Das englische Abstraktum wiederum steht dem im Deutschen nur als technischer Fachbegriff bekannten, lateinischen Spatium nahe. Spatium findet im Deutschen in der Fachsprache der Graphikdesigner Verwendung. Es steht für Abstände zwischen Buchstaben und Zeilen und meint damit auch ein bisschen so etwas wie den Platz, den ein Zeichen haben muss, um erkennbar zu sein. Seinen Ursprung aber hat das Spatium im Schritt, im Sinne von Distanz und Abstand, den ein Schritt schafft; in vielen Sprachen ist das Wort für Schritt ja auch als Maßeinheit bekannt.

Angela Bulloch spricht und denkt, wie ich annehme, zwar einerseits Englisch, andererseits lebt sie seit Jahren in Berlin und spricht und versteht auch Deutsch. In ihren Projekten, gerade der letzten Jahre, begegnen sich auch zwei Gebrauchsweisen des Wortes Raum / Space, die in beiden Sprachen, unter geringfügig unterschiedlichen Bedingungen, auch mit den einander entsprechenden Wörtern möglich sind. Doch werden die je etymologischen und alltagspragmatischen Verunreinigungen nun besonders bedeutsam. Vereinfacht gesagt geht es bei ihr ja zum einen um die dreidimensionale, räumliche Installation und Präsentation von künstlerischen Arbeiten, zum anderen aber um das Thema des Raums in einem Sinne, in dem man auf Deutsch vom Weltraum spricht, gelegentlich im Englischen von Outer Space. Betrifft das erste Thema den Raum als Abstraktum, als Voraussetzung dreidimensionaler künstlerischer Anordnung, so betrifft das zweite einen konkreten Raum oder Nichtraum, den Weltraum. In ihren Arbeiten begegnen sich der Begriff des Raums und ein Raum, der nicht mehr auf den Begriff zu bringen ist.

Bulloch hat immer schon mit Vorstellungen gearbeitet, die das Sichtbare übersteigen. Die Pixel Boxen, die einige Jahre für unterschiedliche Arbeiten eine Art Signature-Oberfläche schufen, waren zugleich eine ständige Aufforderung an die Betrachter, ein nicht sichtbares Mehr mitzudenken. Zum einen waren sie beruhigend schöne, den Proportionen konventioneller Ausstellungsräume wohl entsprechende minimalistische Skulpturen, zum anderen aber bedeuteten sie, anders als minimalistische Skulpturen, nicht sich selbst, sondern minimale Bildpunkte. Sie standen für genau die kleinste Einheit technischer Bilder, die man als solche nicht sieht und sehen soll, wenn einen der Inhalt der Bilder im kulturindustriellen Alltag bannt. Stellt man sich die Bilder zu Bullochs Pixel Boxen vor, hat man nicht nur sehr große, architektursprengende Bilder vor sich. Man hat Bilder vor sich, die aus lauter hochautonomen Einzelobjekten zusammengesetzt sind.

Solche verwirrenden, verunsichernden Gedankenspiele, die man aufgrund von sinnlich wahrnehmbaren Objekten vornimmt, deren Resultate aber bloße Vorstellungen bleiben, gehören bei Kant zu einer wenig beachteten ästhetischen Kategorie. In der Regel wird

English noun, for its part, is close to Latin *spatium,* which only exists as a technical term in German. In German, *Spatium* is a technical term used by graphic designers. It refers to the distances that separate characters and lines and thus describes something a bit like the room *[Platz]* that a sign or character needs to be recognizable. The Latin word *Spatium,* however, has its semantic origins in *pace* or *step,* in the sense of the distance and separation a step creates. Indeed, in many languages, the word for pace or step is also used as a unit of measurement.

I presume that Angela Bulloch speaks and thinks in English; she has lived in Berlin for years, however, and also speaks and understands German. In her projects–especially those of recent years–she combines two different uses of the word *Raum / space,* which are possible in both languages under slightly different conditions. Yet the etymological and everyday pragmatic contaminations now become especially significant. To oversimplify, Bulloch is concerned with the three-dimensional, spatial installation and presentation of artistic works but also with the subject of space, or *Raum,* in the sense of English *outer space* and German *Weltraum.* While the first involves *Raum / space* as an abstraction–a precondition of three-dimensional artistic arrangement–the second involves a concrete space or nonspace: outer space. In Bulloch's works, the concept of space comes together with an actual space that can no longer be conceptualized.

Bulloch has always worked with representations that transcend the visible. The Pixel Boxes that created a kind of signature surface for many of her works for a number of years also constantly challenged viewers to conceive of an *invisible something more.* On the one hand, they were reassuringly beautiful minimalist sculptures that matched the proportions of conventional exhibition spaces well. Unlike minimalist sculptures, however, they did not signify themselves but minimal pixels. They stood for the smallest units of technical images, which cannot and are not supposed to be seen when one is under the spell of the culture industry's images. If one attempts to imagine the images that would correspond to Bulloch's Pixel Boxes, they would not just be very large–large enough to burst the bounds of any architectural container; they would also be made up entirely of highly autonomous individual objects.

For Kant, bewildering and unsettling intellectual games of this kind, which are played on the basis of sensuously perceptible objects but whose results are mere representations, belong to an aesthetic category that receives relatively little attention. Generally speaking, when texts on art refer to Kant's aesthetics, they either discuss the category and aesthetic experience of the beautiful (and hence the artwork), or else they discuss the category of the sublime, by which they usually mean the dynamic sublime (which concerns the aesthetic experience of nature). There is a third category, however, and it is the mathemati-

in Texten über Kunst von seiner Ästhetik entweder die Kategorie des Schönen und der ästhetischen Erfahrung des Schönen (das betrifft das Kunstwerk) oder die Kategorie des Erhabenen, gemeint ist in der Regel das dynamische Erhabene (und das betrifft die ästhetische Naturerfahrung), erwähnt. Die dritte Kategorie aber wäre das mathematische Erhabene: Dies meint die intellektuelle Vorstellung, die jede sinnliche Vorstellung übersteigt, und die wird gerne daran exemplifiziert, dass wir uns auf der Basis mathematischen Denkens Vorstellungen von Unendlichkeit machen, ohne Unendlichkeit sehen zu können – während man beim dynamischen Erhabenen ja durchaus in den Abgrund blickt und eine überwältigende Sinneserfahrung macht. Wahrscheinlich interessiert das mathematische Erhabene die Kunstbetrachtung nicht, weil es ohne besondere Sinneseindrücke auszukommen scheint, während in den beiden anderen Kategorien je ein Sinneseindruck auf die menschlichen Vermögen (Lust / Unlustgefühle, Vernunft und Willen) direkt einwirkt.

Nun wissen wir aber schon lange, dass die Bildende Kunst längst in ein Zeitalter der Metakunst eingetreten ist. Das sinnlich Übersteigende muss nicht notwendig von einer unsinnlich gedachten Symbolverarbeitung ausgelöst sein, ja die Vorstellung einer Auslösung und Ausstattung mentaler Prozesse durch eine extern als Sinnesreiz angebotene Kunst scheint insgesamt fragwürdig. Sind nicht auch immer schon die Vorgänge, die das sinnlich Vorstellbare übersteigen, auf sinnliche Vorstellungen angewiesen? Selbst die strenge Idee des mathematischen Erhabenen, dass ich eine rein vernunftgemäße Vorstellung von Unendlichkeit nur dadurch habe, dass ich mit der liegenden Acht rechne und mir keine endlosen Räume dafür vorstellen muss, kann nicht ignorieren, dass endlose und unübersichtliche Zahlenreihen bereits zum Teil wie Sinnesüberwältigungen wirken.

Auf jeden Fall ist aber für die heutige Situation, in der das mathematische Erhabene auf immer neue Weise in geschriebene wie in gefilmte Science Fiction integriert wurde, eine solche Trennung nicht mehr aufrecht zu erhalten. Ja, die zuständige naturwissenschaftliche Forschung behilft sich selbst mit der Modellierung sinnlicher Anschauung. Wer je Gary Shius computergeneriertes, dreidimensionales Modell einer sechsdimensionalen Geometrie gesehen hat, erlebt haargenau die Mischung aus intellektueller Bewältigung, formaler Denkbarkeit und Unvorstellbarkeit, die das mathematische Erhabene auszeichnet, basierend aber auf prächtigen violetten Modellen, nicht auf Zahlen. Dieses Modell ist nicht die sinnliche Sache selbst, die die Erschütterung auslöst, sie ist wie die Zahl, die mich etwas denken lässt, was ich gerade nicht sehen kann. Nur reicht der Zeichentypus Zahl hier nicht, ein komplexeres Sinnesdatum ist nötig, um mir intellektuell vorzustellen, was ich mir sinnlich nicht vorstellen kann (wie Sechsdimensionalität). Zugleich kann

cal sublime. It refers to an intellectual conception that transcends any possible sensuous experience or representation. An example frequently given is the experience in which we use mathematical thinking to form a representation of infinity, without, however, being able to see infinity, whereas in the case of the dynamic sublime, one actually looks into the abyss and has an overwhelming sensory experience. The reason the mathematical sublime is of so little interest to those who write about art is probably that it appears to make do without particular sensory impressions, whereas each of the other two categories involves a sensory impression that acts directly on the human faculties (feelings of pleasure / displeasure, reason and will).

However, we have long been aware that the visual arts have long since entered an era of meta-art. That which transcends the sensory does not necessarily have to be touched off by an unsensuous, purely intellectual symbolic operation. Indeed, the very notion that an art conceived as external sensory stimulus could touch off and configure mental processes has become questionable. Aren't the processes that transcend sensuous experience or representation always in fact dependent on such representations? Even the rigorous idea of the mathematical sublime—the notion that I only form a purely rational representation of infinity when I operate with the "sideways eight" and strictly avoid imagining infinite spaces—cannot ignore the fact that endless, bewildering rows of numbers already function to some extent as overwhelming sensory experiences.

Today, at any rate, when science fiction writers and filmmakers are constantly finding new ways to incorporate the mathematical sublime, this distinction is no longer tenable. Indeed, even the relevant scientific research relies on the assistance of sensuous models. Anyone who has ever seen Gary Shiu's computer-generated, three-dimensional model of a six-dimensional geometry has experienced precisely that mixture of intellectual mastery, formal conceivability, and unrepresentability that distinguishes the mathematical sublime, in this case, however, based on splendid purple models rather than numbers. This model is not the sensory thing itself that produces the shock. Instead, it is equivalent to numbers, which enable me to conceive of something that I precisely *cannot* see. The difference is simply that in this case, numbers themselves are inadequate. A more complex sensory datum is required, if I am to conceive intellectually of what I cannot conceive of sensuously (for example, six-dimensionality). At the same time, however, these hybrid forms of a sensuously contaminated mathematical sublime are easily distinguished from the dynamic sublime. The latter can only be experienced in a form of existential peril: it must at least be a possibility that its confusing and disorienting, overwhelming form may entail a threat of danger to life and limb. It is really present, not just a phantasm, and the menace it embodies must be overcome by strength of will.

man all diese Mischformen eines sinnlich verunreinigten mathematischen Erhabenen problemlos vom dynamischen Erhabenen unterscheiden: Das kann man nur in einer Form von existenzieller Ausgesetztheit erfahren; es muss wenigstens als Möglichkeit seine unübersichtliche, überwältigende Form mit einer Androhung von Gefahr für Leib und Leben der Betrachter verbinden. Es ist real präsent, kein Phantasma, und seine Drohung kann durch Willenskraft gezähmt werden.

Zwar haben sich Bildende Kunst, Oper und vor allem die Kulturindustrie an Jahrmarkteffekten orientiert, sich immer wieder um die Dimension des dynamischen Erhabenen bemüht und dafür vor allem im Action-Kino eine Überwältigungsästhetik entwickelt, die ihre Wirkung immer nur durch neue Verschärfungszyklen aufrecht erhalten kann. Aber gerade diese Überwältigungsästhetik ist nur als Erinnerung an die existenzielle Erhabenheitserfahrung angelegt und ist ganz besonders an die Bequemlichkeit des Kino-Mobiliars und die Geborgenheit im Kollektiv gebunden.

Die Engführung aus einer Diskussion des Raums der künstlerischen Installation und des Raums im Sinne von Weltraum, die Angela Bulloch betreibt, ist auch davon ganz entschieden inspiriert, dass es für beide Begriffe von Raum keine Sicherheit und Gewohnheit gibt. Während die Black Box des Kinos in ihren verschiedenen Varianten gute Dienste geleistet hat, die Konstruiertheit der Kino-Erfahrung nicht in das reflexive Bewusstsein des Publikums dringen zu lassen, so sind im Gegensatz dazu im Raum der Installation der Bildenden Kunst trotz aller Standardisierungsbemühungen im Prinzip alle Maße möglich. Gerade in letzter Zeit hat sich eine entsprechende Thematisierung des Ausstellungsraums radikalisiert. Darüber hinaus spiegelt sich die allgemeine Wertkrise der Bildenden Kunst, die keine Maße zu kennen scheint, keine Grenze nach oben, aber zugleich keinen anderen außermonetären Maßstab weiß, der dem Markt entgegenzusetzen wäre, auf unheimliche Weise in der Standardlosigkeit der künstlerischen Räume.

Bulloch nutzt diese allgemeine Unsicherheit für ein doppeltes Manöver aus Verunsicherung und Setzung. Ihre Arbeiten betonen zum einen die prinzipielle Ungeklärtheit (im Sinne aufgelöster oder unbegründeter, rein konventioneller Standards) oder auch Offenheit der Raumfrage, indem sie monströse Maße inhaltlich oder formal abrufen (Riesenpixel, Verweise auf unvorstellbare Distanzen etc.); zum anderen aber sind sie selbst durch sehr klare Konturen gekennzeichnet. Sie expandieren nicht, verändern sich nicht, jedenfalls nicht außerhalb einer immanenten Stochastik (etwa bei den zufallsgenerierten Farbwechseln der Würfel); sie sind formal und konzeptuell wohldefiniert. Sie betonen, dass sie im Prinzip keinen Grund, keine Übereinkunft haben, auf die sie sich beziehen könnten. Doch im nächsten Schritt offerieren sie die Immanenz der Arbeit selbst als

True, the visual arts, opera, and above all the culture industry, oriented as they are around funhouse effects, are constantly striving for the dimension of the dynamic sublime. Action movies in particular have developed an aesthetic of "overwhelming force" that can only maintain its effectiveness through constant escalation. Precisely this aesthetic, however, is only designed as a reminder of the actual existential experience of the sublime, and it is very closely tied to the comfort of the movie theater and the feeling of safety and security in the collective.

Angela Bulloch's close intertwining of a discussion of the space of artistic installation and a discussion of space in the sense of outer space is also very much inspired by the fact that neither of these concepts of space involves any certainty or security, any habits. The black box of the movie theater in its different variants has done a good job of preventing the constructedness of the movie-theater experience from entering the consciousness of the audience. By contrast, in the space of the visual art installation, despite all efforts at standardization, any dimensions are still theoretically possible. Just recently, this aspect of exhibition space has been thematized in a particularly radical manner. Moreover, the general crisis of value in the visual arts, which seems to have no sense of proportion, no upper limit, but also seems to have no non-monetary criterion that might be set against the market, is strangely reflected in the absence of standards for artistic spaces.

Bulloch utilizes this general uncertainty for a dual maneuver that combines unsettling gestures with forthright assertion. On the one hand, her works emphasize the fundamentally unresolved character (in the sense of abrogated or unfounded, purely conventional standards) and openness of the space question by evoking monstrous dimensions (giant pixels, references to unimaginable distances, etc.). On the other hand, they themselves are distinguished by very clear outlines. They do not expand, and they do not change, at least not outside an immanent range of possible variation (as, for example, with cubes that change color randomly). Yet in the very next moment, they offer the immanence of the work itself as certainty and security. They relate to the universe as tragedy does to the pantheon. They are closed, self-contained, and completely thought through, even though they present themselves as instances of what cannot be closed and cannot be completely thought through.

This becomes especially clear in Bulloch's series of literal views of the cosmos. With the help of a computer program that is by no means especially esoteric, the artist has worked out views of the universe that are not oriented around the vantage point of the earth-dweller. As a rule, the maps of the universe with which we are familiar tend to depict a section of the cosmos as a portion of the night sky seen from a particular point on earth. The child's classic question when confronted with a map—where are we?—cannot really be answered

Sicherheit. Zum Weltall verhalten sie sich wie die Tragödie zum Götterhimmel. Sie sind abgeschlossen und zu Ende gedacht, auch wenn sie sich als ein Fall dessen ausgeben, das nicht zu Ende gedacht und abgeschlossen werden kann.

Am deutlichsten wird das in der Serie von buchstäblichen Ansichten des Kosmos. Mit Hilfe eines keineswegs esoterischen Computerprogramms hat Bulloch Ansichten des Weltalls ausgerechnet, die sich nicht am Standpunkt des Erdbewohners orientieren. Karten des Weltalls sind in der Regel als solche bekannt, die vom Nachthimmel von einem bestimmten Punkt der Erde aus gesehen einen Ausschnitt des Kosmos zweidimensional darstellen. Die klassische Kinderfrage an Landkarten – wo sind wir? – lässt sich für Karten des Weltalls nicht richtig beantworten. Wir sind auf der Karte nicht eingezeichnet. Wir sind die hinter dem Fernrohr. Bullochs Ansichten aber zeigen auch Blicke von woanders her, nun sind wir nicht einmal mehr hinter dem Fernrohr. Menschliche Subjektivität ist an keinem Ende an ihrem Zustandekommen beteiligt, nur die alten menschlichen Genres Landschaft und Landkarte geben noch einen Rahmen vor. In einem Fall hat Bulloch sich den lakonischen Scherz erlaubt, als Ausgangspunkt den Planeten Gleise 581 C im Sternbild Waage, der den roten Zwerg Gleise 581 umkreist, auszuwählen. Auf Gleise 581 C sollen Temperaturen zwischen –3 und 40 Grad Celsius herrschen und ein Leben wie auf der Erde möglich sein. In anderen Fällen führen die Blicke ganz an uns vorbei.

Ein so verschobenes Bild des Universums leistet aber keinen Beitrag zur Astronomie, es trägt vielmehr etwas zu einer möglichen kopernikanischen Wende im Feld des Sehens in der Bildenden Kunst bei. Die Räume, in denen künstlerische Projekte stattfinden, werden heutzutage nicht mehr von den Kuben und anderen praktisch sinnvollen Formen der Architektur zwingend bestimmt. Da sie praktischen Erwägungen nicht gehorchen müssen, ist alles, was berechenbar ist, auch als Display, Rahmen, Kontainer für Kunst vorstellbar – wodurch aber dieser Rahmen auch jede „Sache selbst" überwuchert. Bulloch versucht darum nicht, trotz ihres entwickelten Interesses für virtuelle Architektur, tatsächlich diese neuen Möglichkeiten auszureizen. Sie entwickelt vielmehr Modelle für die Probleme, die sich in dieser neuen Situation ergeben, in der die Räume der Kunst nicht mehr von den Zimmern abgeleitet werden können, denen die Kunstwerke einmal die Vorstellung einer Außenwelt liefern sollten.

Diese Modelle sind keine konkrete Meta-Architektur, sondern eher konzeptuelle Szenarios für folgende grundsätzliche Frage: Da die Trennung zwischen dem abstrakten Begriff des Raums – als a priori von Vorstellbarkeit – und dem konkreten, sozusagen dimensional ungeklärten „Raum" Universum verschwimmt, müssen Astronomen quasi ohne a priori auskommen oder mit sehr viel weiteren, ungenaueren oder unanschaulicheren a priori arbeiten als es Raum und Zeit einst waren. Immer wieder gibt es Entdeckungen, die

for maps of the universe. We are not on the map; we are the ones behind the telescope. Bulloch's views, however, also show the universe as seen from elsewhere; now we are not even behind the telescope any longer. Human subjectivity is not involved in the production of these views at either end. Only the old human genres of landscape and land map continue to provide a frame. In one instance, Bulloch permitted herself the laconic joke of choosing the planet Gleise 581 C in the constellation Libra, which orbits the red dwarf Gleise 581, as the vantage point for one of her views. Temperatures on Gleise 581 C are believed to range between 27 and 104 degrees Fahrenheit, and a life like that on earth is believed to be possible there. In other cases, Bulloch's views leave us completely out of the picture.

But such a dislocated image of the universe does not contribute in any way to astronomy. Rather, it contributes to a Copernican revolution in the field of seeing in the visual arts. The spaces in which artistic projects take place today are no longer dictated by cubes or other architectural forms that have practical meanings. Because they do not have to conform to any practical considerations, everything that can be calculated can also be conceived as a display, frame, or container for art. The result, however, is that this frame now overwhelms every possible "thing itself." For this reason—and despite her highly developed interest in virtual architecture—Bulloch does not attempt to take full advantage of these new possibilities. Rather, she develops models for the problems that arise from this new situation, in which the spaces of art can no longer be derived from the rooms to which artworks were formerly supposed to furnish the representation of an outside world.

These models are not concrete meta-architecture but conceptual scenarios for this fundamental question: since we are witnessing the blurring of the distinction between the abstract concept of space—as an a priori condition of human experience and conception—and the concrete, as it were dimensionally unresolved "space" of the universe, astronomers are forced in a sense to make do without a prioris or to work with a prioris that are substantially broader, vaguer, and less vivid than space and time. Again and again, discoveries are made that do not just produce a new fact but revise and destabilize the categories themselves and ultimately the a prioris on which they are based. Formally, this problem is similar to that which arises in the new, open spatiality of the visual arts. Between Prada Store, experiential installations, and Archigram-inspired creations on the practical side, and art as its own space and *spatial* effect as the new *artistic* effect on the theoretical side, installation art is entering a realm in which every individual project must determine afresh against what wall, what external environment, and even what kind of external environment it wishes to stand out. Theoretical determinations that are more familiar from the conceptual avant-garde—determinations of art's relationship to concrete sites and

nicht nur ein neues Faktum hervorbringen, sondern auch die Kategorien und schließlich die a priori, auf denen die Kategorien basieren, revidieren oder erschüttern. Dies ist formal ein ähnliches Problem, wie es in der neu geöffneten Räumlichkeit der Bildenden Kunst auftaucht: Zwischen einerseits Prada Store, Erlebnis-Installationen und Archigram-Nachempfindungen auf der praktischen Seite und andererseits Kunst als ihr eigener Raum und Raumwirkung als Nachfolgerin der Kunstwirkung auf der theoretischen Seite gerät Installations-Kunst in ein Gebiet, wo jedes einzelne Projekt aufs Neue bestimmen muss, im Gegensatz zu welcher Wand, welchem Außen und darüber hinaus welcher Art von Außen es sich absetzen will. Die eher aus der konzeptuellen Avantgarde bekannten theoretischen Bestimmungen des Verhältnisses von Kunst zu konkreten Orten und den Grenzen und Öffnungen ihrer Institutionen kehrten auf eine technisch praktische Ebene zurück: in Bezug auf die architektonisch-mathematische Seite konkreter Räume, zunächst nicht auf deren politische Seite.

Die Verbindung von geschlossener Räumlichkeit und der Struktur des Kosmos führt in Platons *Timaios* zum Konzept der vollkommenen Körper, Platonische Körper genannt. Diese sind dem Verhältnis zwischen der Menge der vier Elemente im Universum entsprechend konstruierte geometrische Körper. Bulloch rekonstruiert diese Körper mit Lichtumrissen und setzt sie so in Beziehung zu den chaotischen Sternbildern, die ja auch durch gedachte Linien eine Form bekamen, die man sich merken konnte. Die Platonischen Körper galten aber als Bauplan für das Universum selbst. Die Beschäftigung mit der Geometrie galt als unmittelbare Versenkung in die Struktur des Universums. Die quasi ursprüngliche Vermittlung zwischen Zahlen und dem Unendlichen in der Anschauung waren die geometrischen Körper und die von ihnen ausgehende Vorstellung von Räumlichkeit. In dem Moment, wo solche Vorstellungen nicht mehr herstellbar sind, beginnt die Relation des mathematischen Erhabenen. Und damit die Moderne.

In der Moderne haben aber einfache arbiträre Zeichen (Zahlen und Buchstaben) zwischen undarstellbarem Kosmos und menschlicher Vernunft vermitteln können. Bulloch stellt sich nun der Tatsache, dass das in der Astrophysik des 20. Jahrhunderts nicht mehr möglich ist oder nur unter enormen Schwierigkeiten und Vermittlungen zwischen formalen Sprachen und menschlichen Vorstellungsvermögen und Kapazität. In dieser Situation wird die für die Herstellung von Anschauungen und deren Kritik zuständige Kunst, so könnte man Bulloch interpretieren, mit der Aufgabe betraut, Modelle herzustellen, aber nicht mehr in der Funktion, die sie zu Platons Zeiten hatten – als authentische mikrokosmische Entsprechungen des Kosmos – sondern so wie mathematische Zeichen, aber als ein anderer, erweiterter, unreiner Zeichentyp, quasi als Synthese zwischen Platonischer Kosmologie und mathematischem Erhabenem.

to the social and political boundaries and openings of its institutions—are now returning on a practical, technical level: with reference to the architectonic and mathematical aspect of concrete spaces, not at first with reference to their political side.

In Plato's *Timaeus,* the connection between closed, self-contained spatiality and the structure of the cosmos leads to the concept of perfect solids, called Platonic solids. These are geometric solids constructed in accordance with the relationship among the quantities of the four elements in the universe. Bulloch reconstructs them using light to sketch their outlines. In this way, she sets them in relation to the chaotic constellations, which also acquired memorable forms through lines that only existed in thought. The Platonic solids, however, were considered the blueprint of the universe itself. The study of geometry was regarded as a form of direct immersion in the structure of the universe. Geometric solids and the representation of spatiality that was based on them functioned as the almost primordial mediation between numbers and the infinite in perception. The moment when such representations cease to be possible is the beginning of the mathematical sublime, and with it of modernity.

In modernity, however, simple, arbitrary signs (numbers and letters) were able to mediate between an unimaginable cosmos and human reason. Bulloch confronts the fact that in the astrophysics of the twentieth century, this is no longer possible, or if it is, then only with enormous difficulties and mediations between formal languages and human imaginative faculties. One interpretation of Bulloch would be to say that in this situation, art, which is responsible for producing and critiquing perceptions, is entrusted with the task of producing models. These models, however, no longer have the function they did in Plato's time, as authentic, microcosmic analogies of the cosmos. Instead, they resemble mathematical signs, but as a different, expanded, impure type of sign, almost a kind of synthesis of Platonic cosmology and the mathematical sublime.

At the same time, however, Bulloch ties this practice to user- and observer-related stagings. She delineates a phenomenology of the crisis of space that—from early works (*Flyover Genoa* and *Fly Over Genoa*) to current works (*Vanishing Waiting Room*)—consistently ties it to the presence of social beings and their interaction with space. True, the current crisis of space is not simply a social crisis. It cannot be reduced to the thesis (correct though it certainly is) that a complex and confusing tunnel is only a space of fear as long as there is also the social phantasm that one is afraid of encountering there. It is also the crisis of a world that has become so accustomed to the standard (standard space) as orientation—from which deviations can always be staged and which provides a stable background for a foreground—that it is deeply shaken when we suddenly meet prototypes everywhere, in art and astrophysics. It is a crisis in which every work and every corner of the cosmos

Zugleich bindet Bulloch aber diese Praktik an benutzer- und betrachterbezogene Inszenierungen. Sie entwirft eine Phänomenologie der Raumkrise, die von frühen Arbeiten (*Flyover Genoa* und *Fly Over Genoa*) bis zu aktuellen (*Vanishing Waiting Room*) darin konsistent ist, dass sie sie an die Anwesenheit sozialer Wesen und deren Umgang mit Raum bindet. Die Raumkrise von heute ist zwar nicht einfach nur eine soziale Krise. Sie lässt sich nicht auf den sicher richtigen Befund reduzieren, dass ein unübersichtlicher Tunnel nur dann ein Angstraum ist, wenn es das soziale Phantasma gibt, das man an den uneinsehbaren Stellen erwarten könnte. Es ist auch die Krise einer Welt, die sich so an den Standard (Standardraum) als Orientierung gewöhnt hat, von dem man immer Abweichungen inszenieren kann und der einen stabilen Hintergrund für einen Vordergrund liefert, dass sie erschüttert wird, wenn wir überall, in der Kunst wie in der Astrophysik mit Prototypen zu tun haben. Bei der jedes Werk und jede Ecke des Kosmos wie eine Singularität zu behandeln ist, die sowohl eine Regel formuliert als auch deren erster und womöglich einziger Fall ist. Doch ist die Angst vor der absoluten Singularität – der elften Dimension, dem Bewohner des schwarzen Lochs, dem absolut anderen U-Bahn-Passagier –, die hinter jeder Ecke lauern könnte, auch ein zuallererst soziales Phänomen, eine Art radikalisierte Xenophobie, die ziemlich gezielt ihre kosmischen und künstlerischen Allegorien gesucht hat. Angela Bullochs aufklärerischer Verdienst ist es, in großer Heiterkeit und mit großer Gelassenheit diese Struktur als solche kenntlich zu machen – den Verlust dieser Übersicht überaus übersichtlich zu gestalten.

must be treated as a singularity that both formulates a rule and is also that rule's first and perhaps only instance. Yet the fear of the absolute singularity—the eleventh dimension, the inhabitant of the black hole, the absolutely other subway passenger—that could be lurking around every corner is also and above all a social phenomenon, a kind of radicalized xenophobia that has sought out its cosmic and artistic allegories quite single-mindedly. It is Angela Bulloch's great merit to have advanced the cause of enlightenment by lending—with great humor and great composure—a very clear form to this unclarity.

Translated from the German by James Gussen

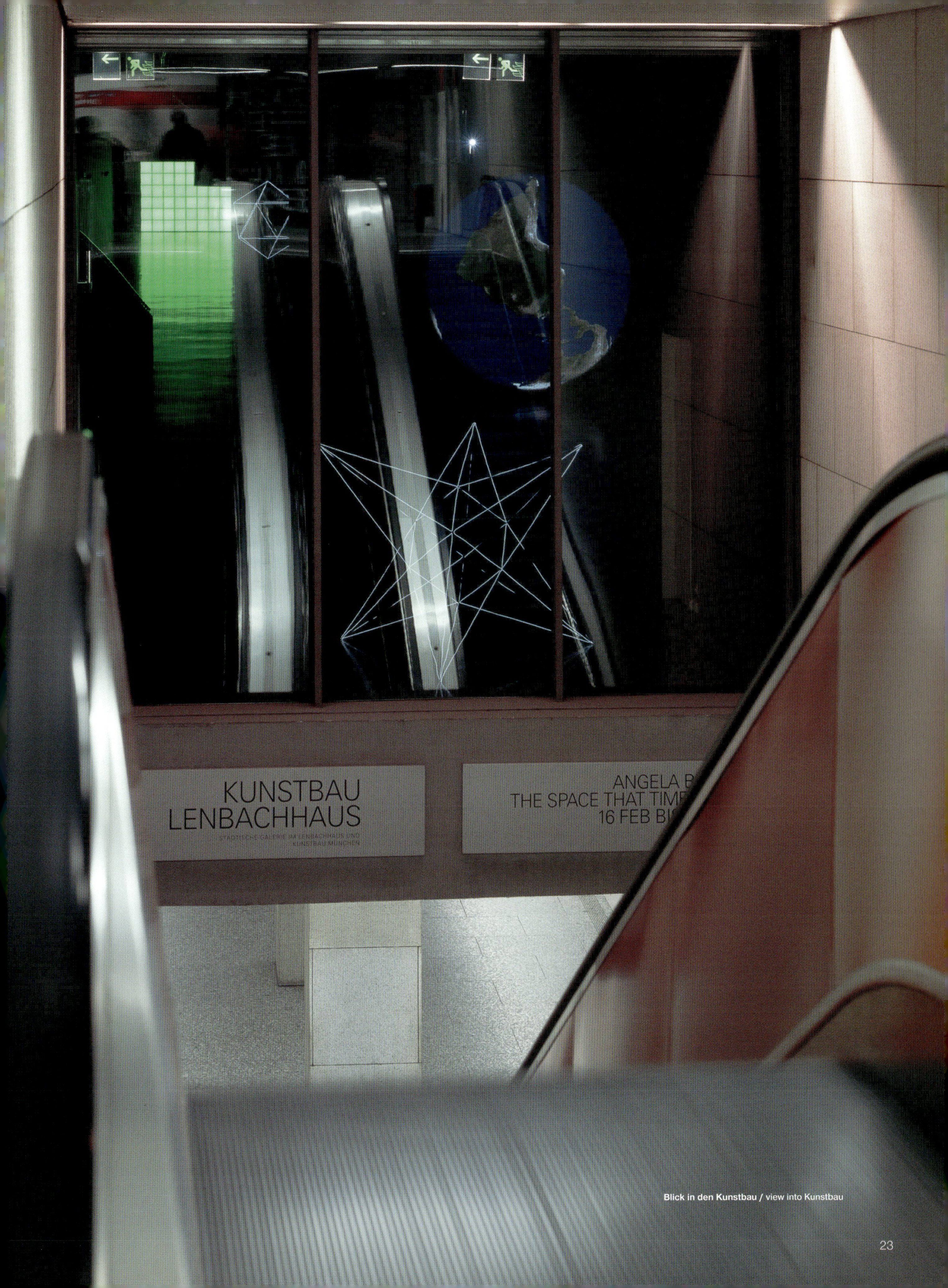

Blick in den Kunstbau / view into Kunstbau

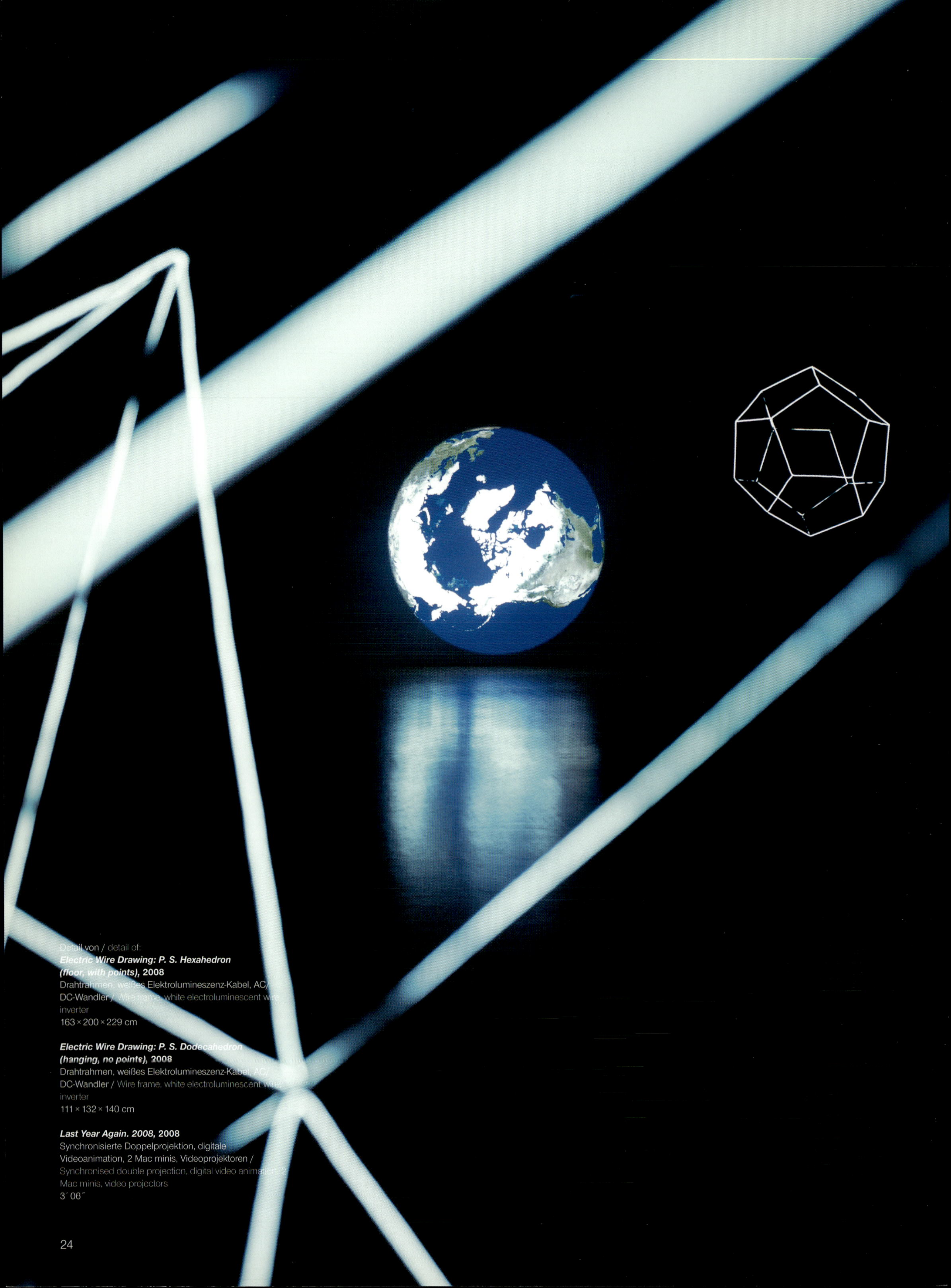

Detail von / detail of:
***Electric Wire Drawing: P. S. Hexahedron (floor, with points)*, 2008**
Drahtrahmen, weißes Elektrolumineszenz-Kabel, AC/DC-Wandler / Wire frame, white electroluminescent wire, inverter
163 × 200 × 229 cm

***Electric Wire Drawing: P. S. Dodecahedron (hanging, no points)*, 2008**
Drahtrahmen, weißes Elektrolumineszenz-Kabel, AC/DC-Wandler / Wire frame, white electroluminescent wire, inverter
111 × 132 × 140 cm

***Last Year Again. 2008*, 2008**
Synchronisierte Doppelprojektion, digitale Videoanimation, 2 Mac minis, Videoprojektoren / Synchronised double projection, digital video animation, 2 Mac minis, video projectors
3′ 06″

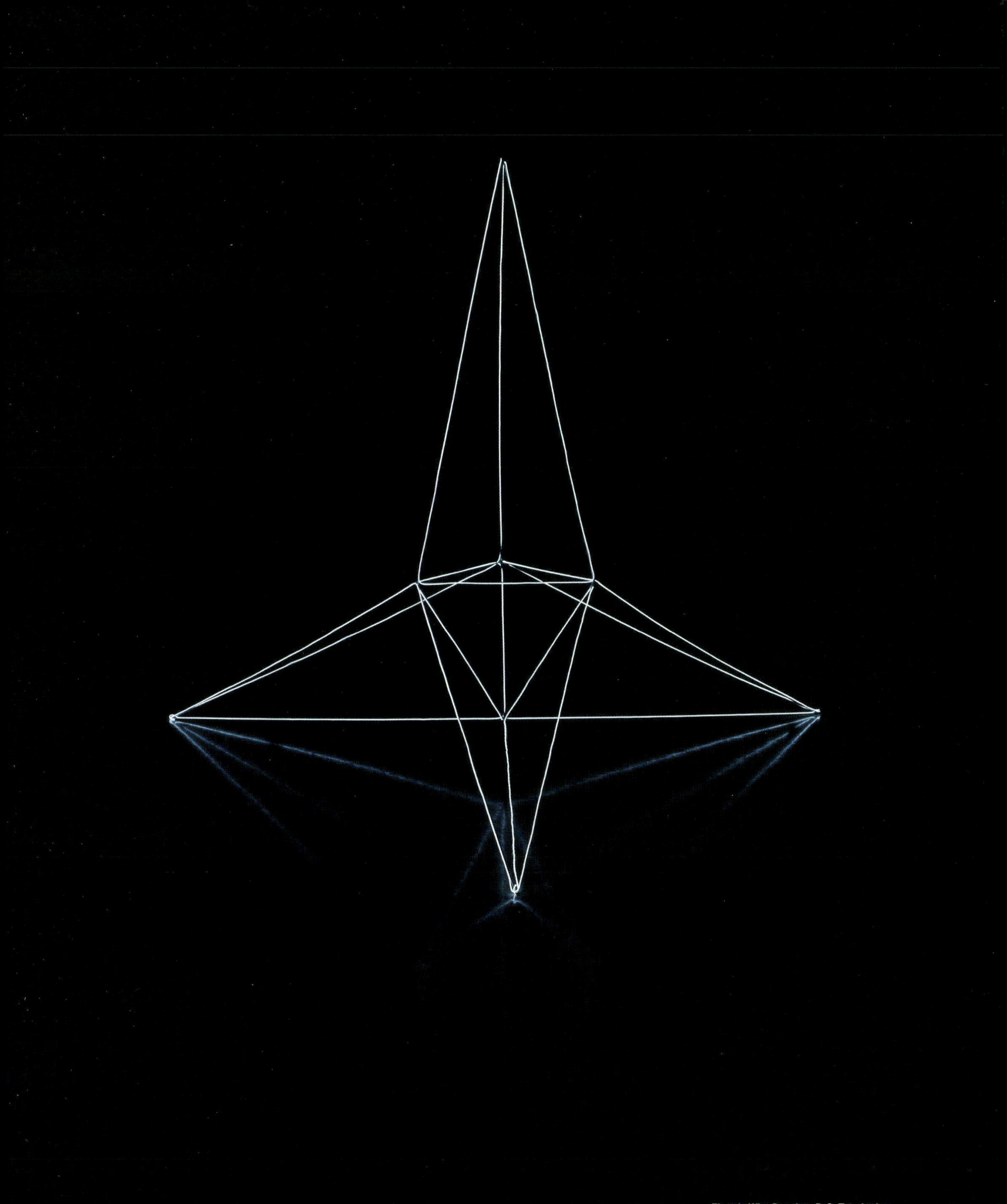

***Electric Wire Drawing: P. S. Tetrahedron (floor, with points)*, 2008**
Drahtrahmen, weißes Elektrolumineszenz-Kabel, AC/DC-Wandler / Wire frame, white electroluminescent wire, inverter
170 × 182 × 211 cm

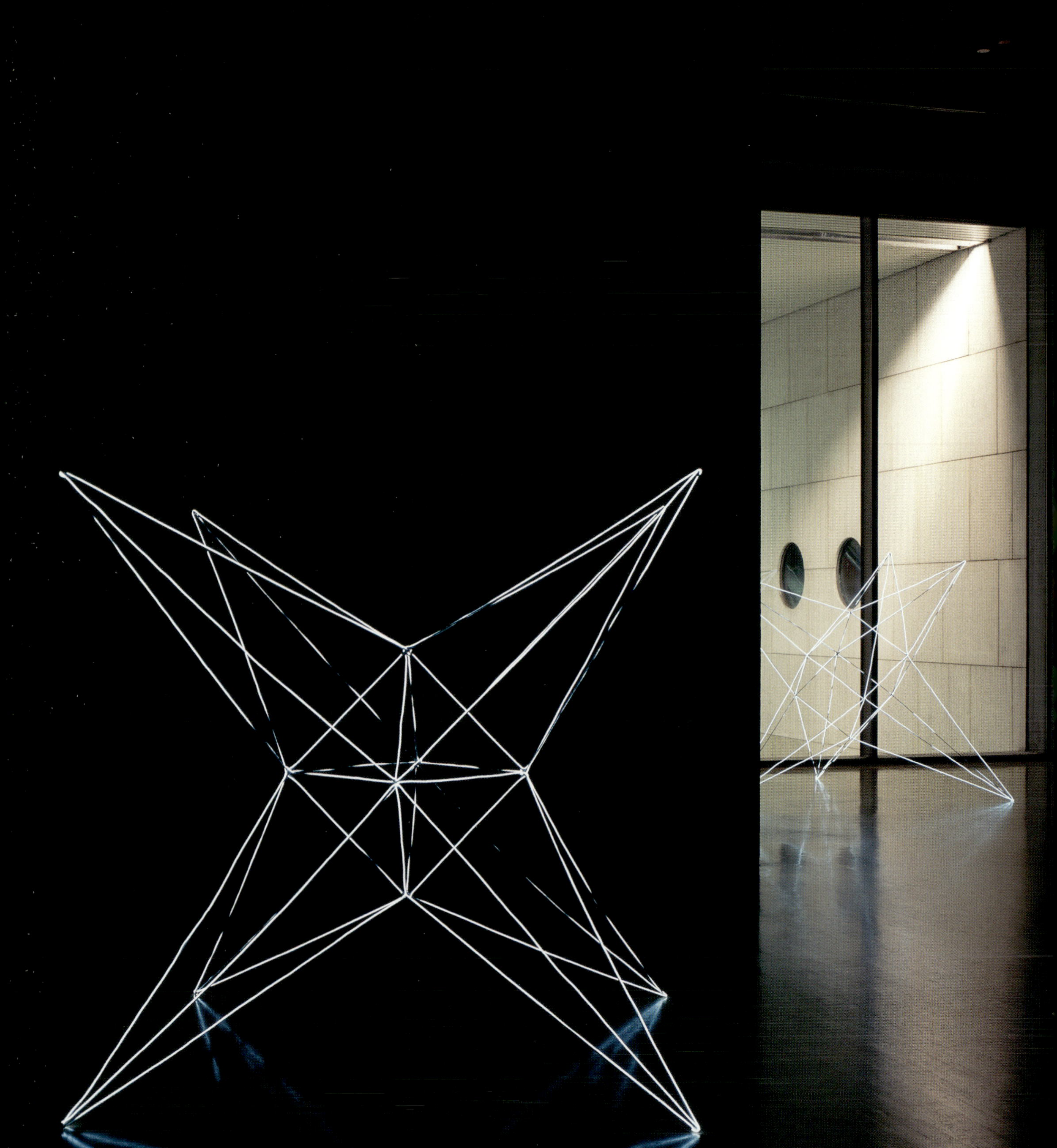

***Electric Wire Drawing: P. S. Octahedron (floor, with points)*, 2008**
Drahtrahmen, weißes Elektrolumineszenz-Kabel, AC/DC-Wandler / Wire frame, white electroluminescent wire, inverter
159 × 225 × 225 cm

***Electric Wire Drawing: P. S. Hexahedron (floor, with points)*, 2008**
Drahtrahmen, weißes Elektrolumineszenz-Kabel, AC/DC-Wandler / Wire frame, white electroluminescent wire, inverter
163 × 200 × 229 cm

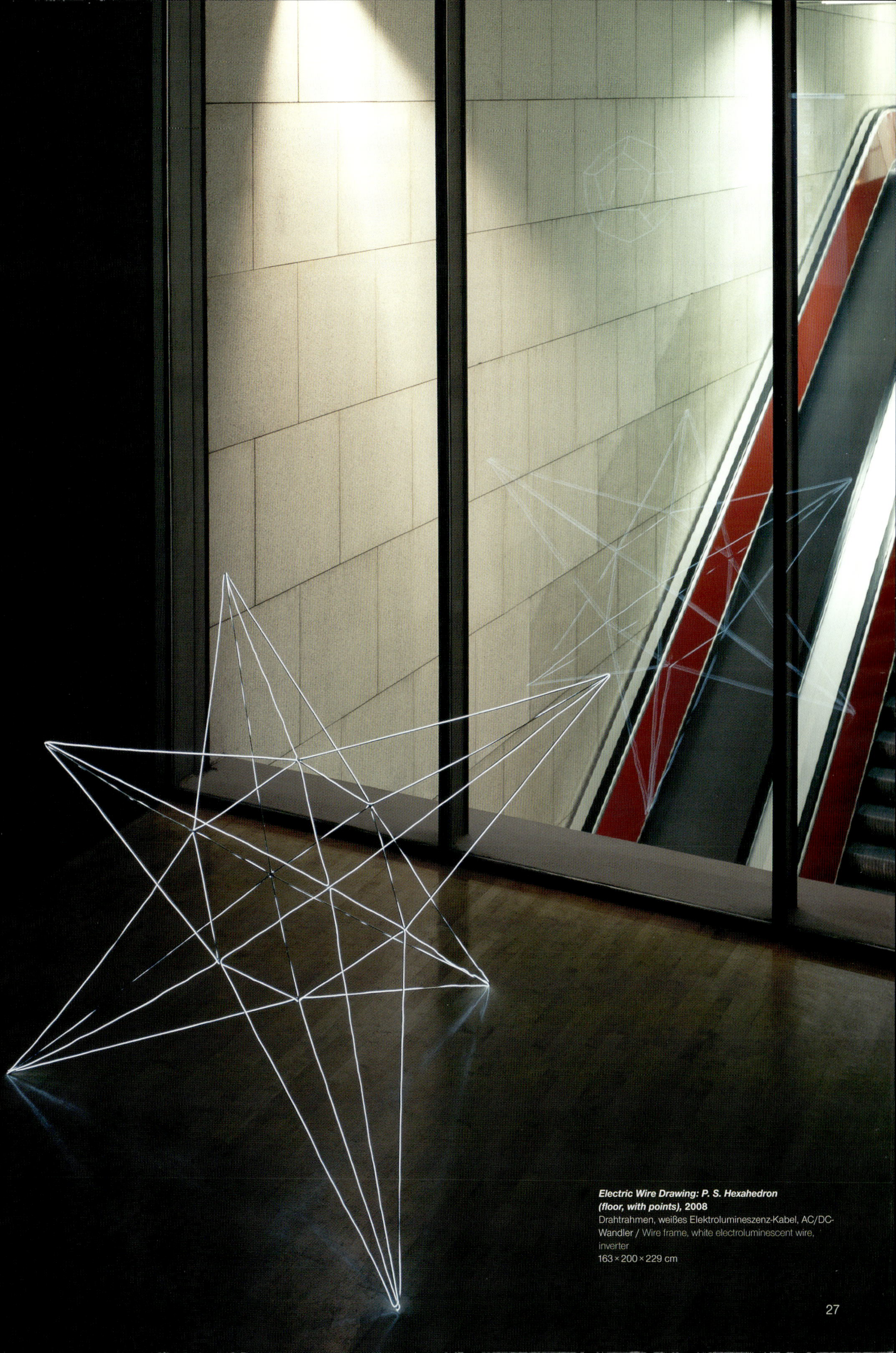

***Electric Wire Drawing: P. S. Hexahedron (floor, with points)*, 2008**
Drahtrahmen, weißes Elektrolumineszenz-Kabel, AC/DC-Wandler / Wire frame, white electroluminescent wire, inverter
163 × 200 × 229 cm

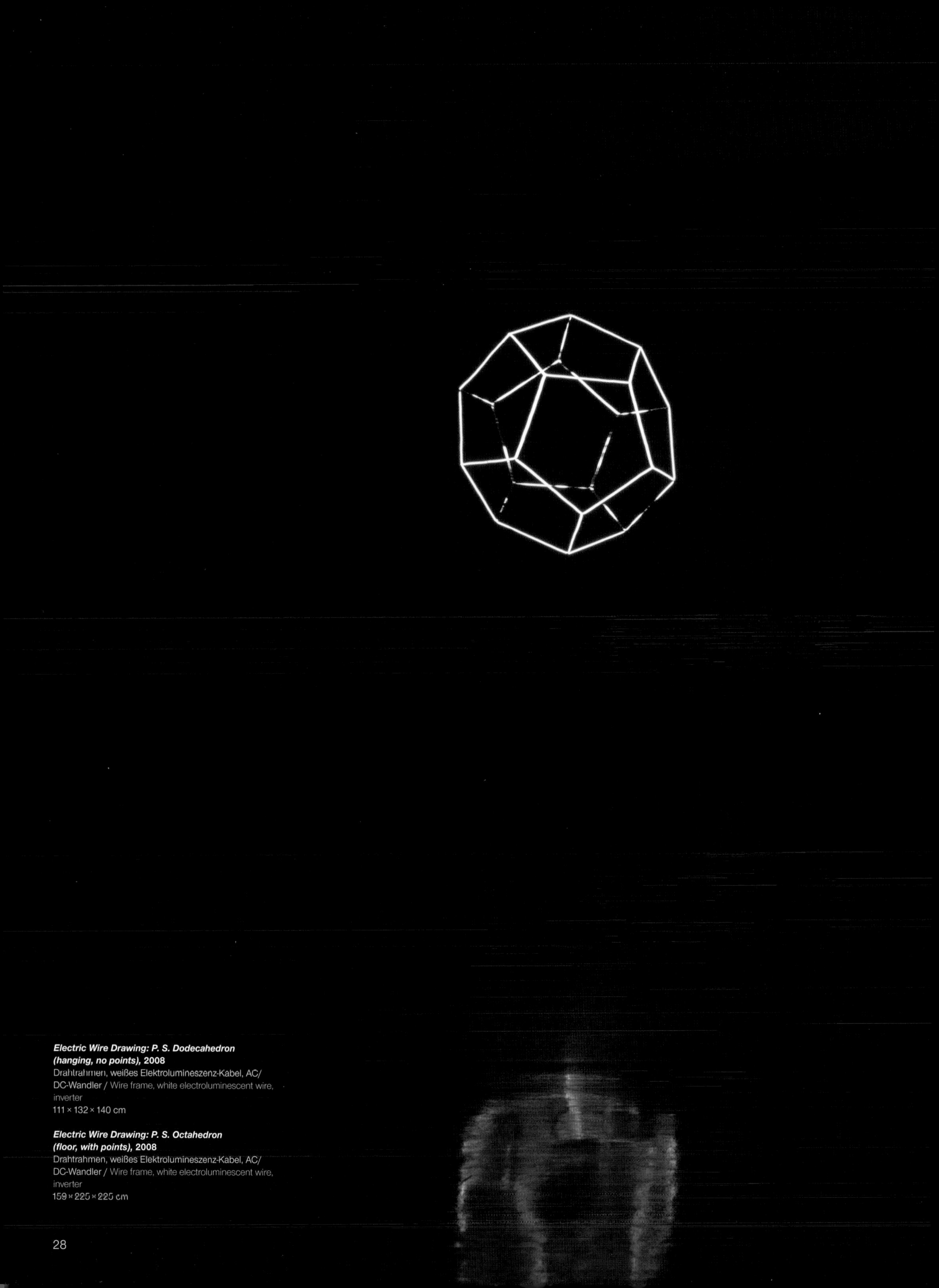

***Electric Wire Drawing: P. S. Dodecahedron (hanging, no points)*, 2008**
Drahtrahmen, weißes Elektrolumineszenz-Kabel, AC/DC-Wandler / Wire frame, white electroluminescent wire, inverter
111 × 132 × 140 cm

***Electric Wire Drawing: P. S. Octahedron (floor, with points)*, 2008**
Drahtrahmen, weißes Elektrolumineszenz-Kabel, AC/DC-Wandler / Wire frame, white electroluminescent wire, inverter
159 × 225 × 225 cm

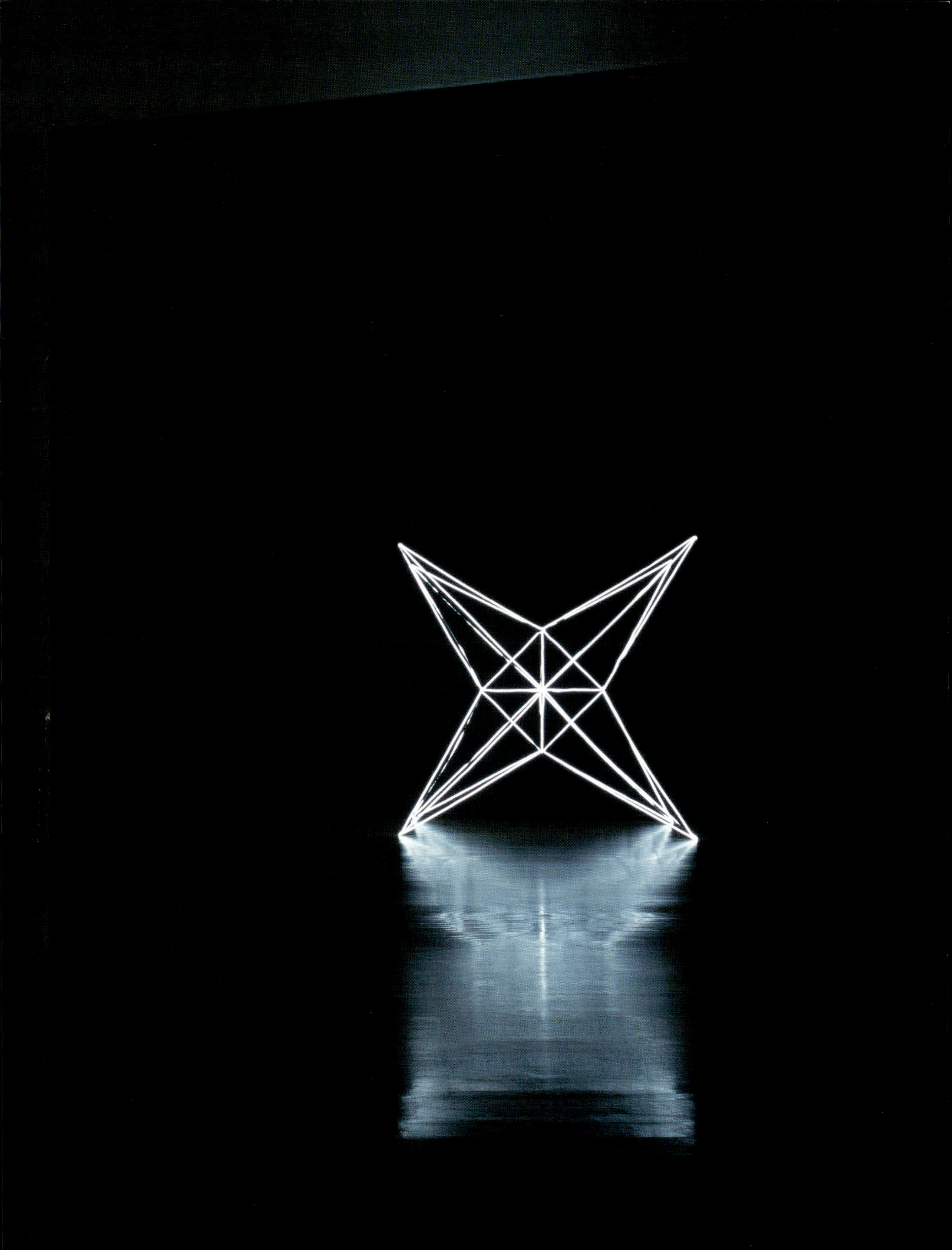

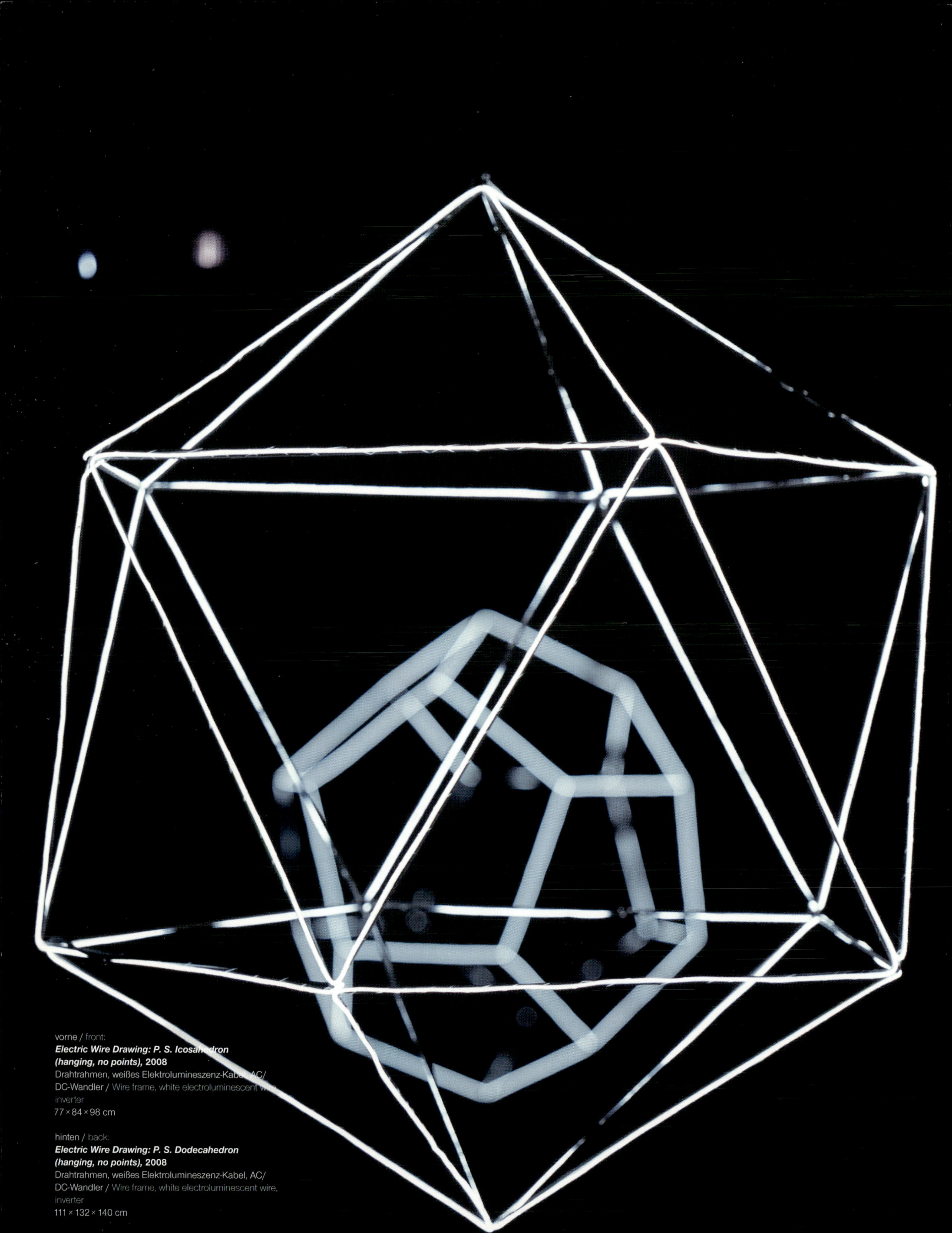

vorne / front:
***Electric Wire Drawing: P. S. Icosahedron (hanging, no points),* 2008**
Drahtrahmen, weißes Elektrolumineszenz-Kabel, AC/DC-Wandler / Wire frame, white electroluminescent wire, inverter
77 × 84 × 98 cm

hinten / back:
***Electric Wire Drawing: P. S. Dodecahedron (hanging, no points),* 2008**
Drahtrahmen, weißes Elektrolumineszenz-Kabel, AC/DC-Wandler / Wire frame, white electroluminescent wire, inverter
111 × 132 × 140 cm

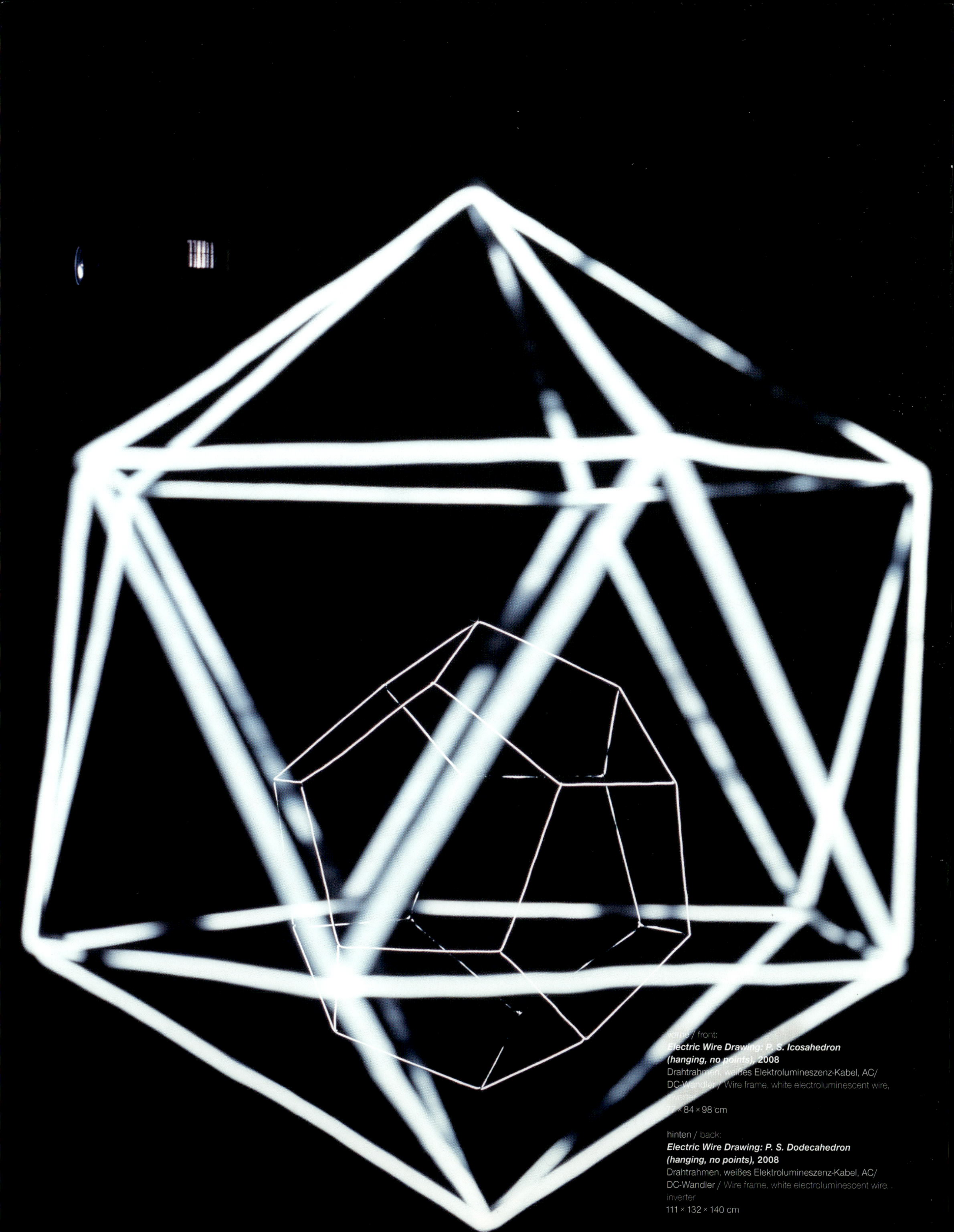

vorne / front:
***Electric Wire Drawing: P. S. Icosahedron (hanging, no points)*, 2008**
Drahtrahmen, weißes Elektrolumineszenz-Kabel, AC/DC-Wandler / Wire frame, white electroluminescent wire, inverter
77 × 84 × 98 cm

hinten / back:
***Electric Wire Drawing: P. S. Dodecahedron (hanging, no points)*, 2008**
Drahtrahmen, weißes Elektrolumineszenz-Kabel, AC/DC-Wandler / Wire frame, white electroluminescent wire, inverter
111 × 132 × 140 cm

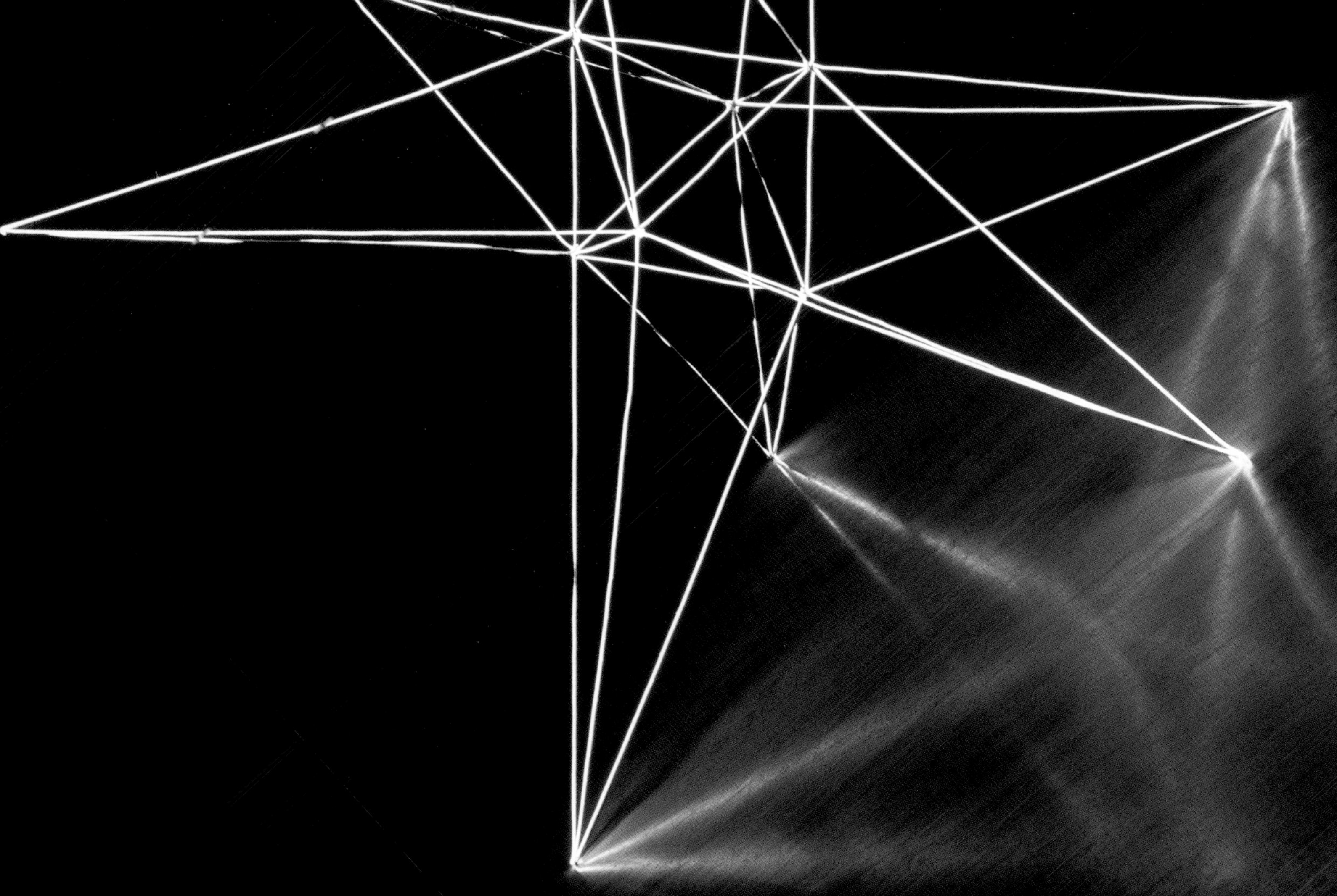

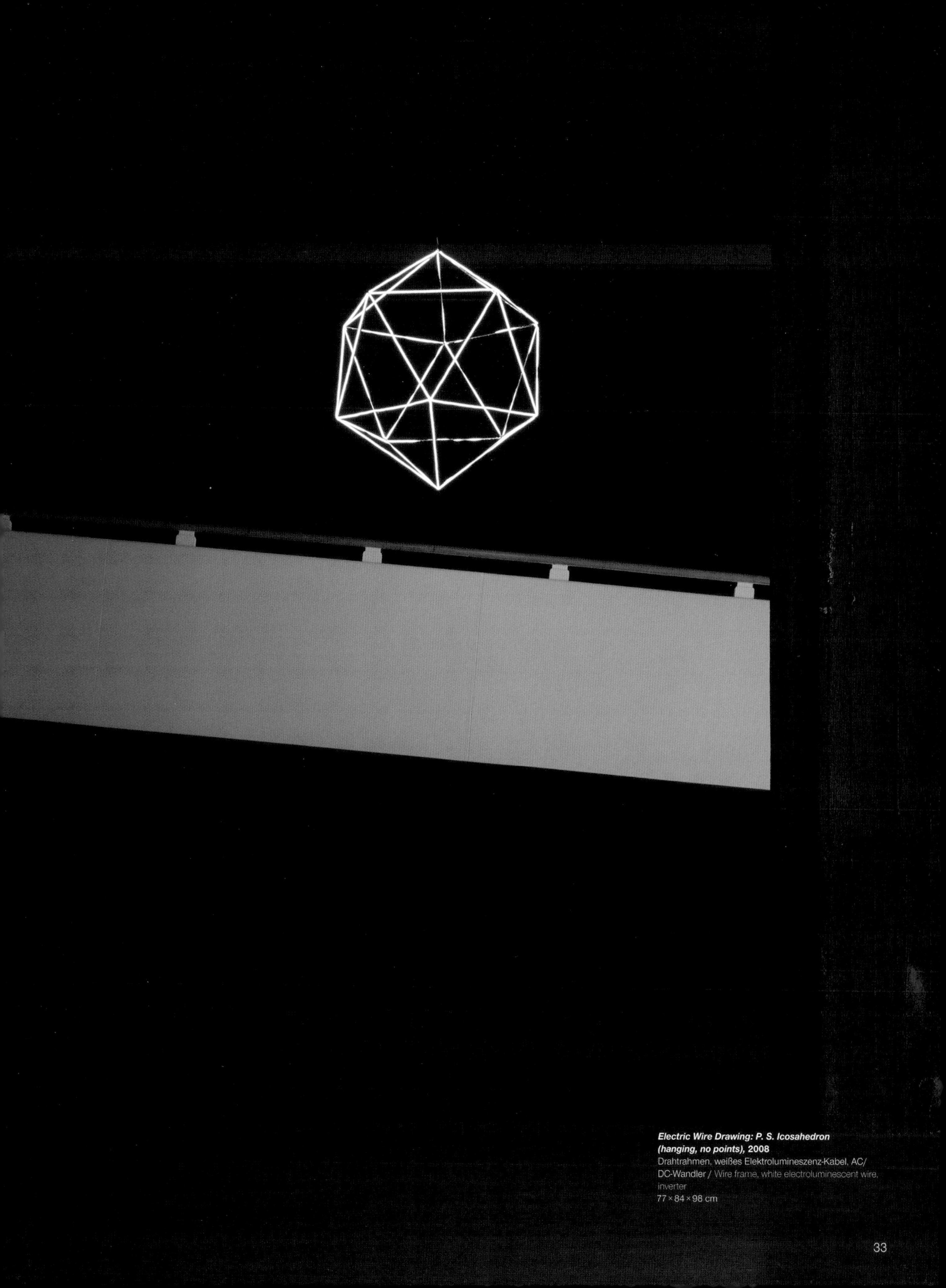

***Electric Wire Drawing: P. S. Icosahedron (hanging, no points)*, 2008**
Drahtrahmen, weißes Elektrolumineszenz-Kabel, AC/DC-Wandler / Wire frame, white electroluminescent wire, inverter
77 × 84 × 98 cm

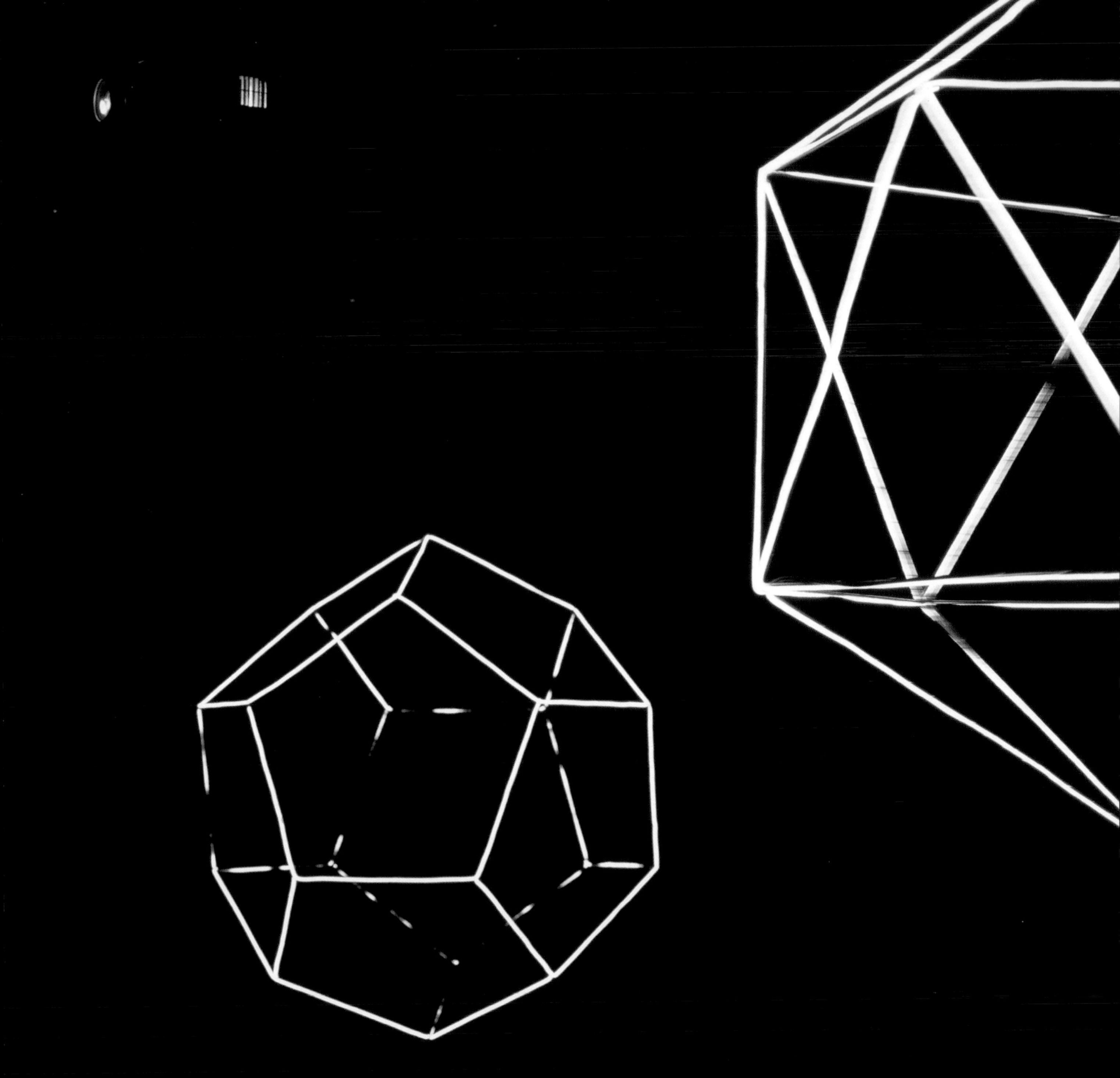

***Electric Wire Drawing: P. S. Dodecahedron (hanging, no points),* 2008**
Drahtrahmen, weißes Elektrolumineszenz-Kabel, AC/DC-Wandler / Wire frame, white electroluminescent wire, inverter
111 × 132 × 140 cm

***Electric Wire Drawing: P. S. Icosahedron (hanging, no points),* 2008**
Drahtrahmen, weißes Elektrolumineszenz-Kabel, AC/DC-Wandler / Wire frame, white electroluminescent wire, inverter
77 × 84 × 98 cm

Electric Wire Drawing: P. S. Octahedron

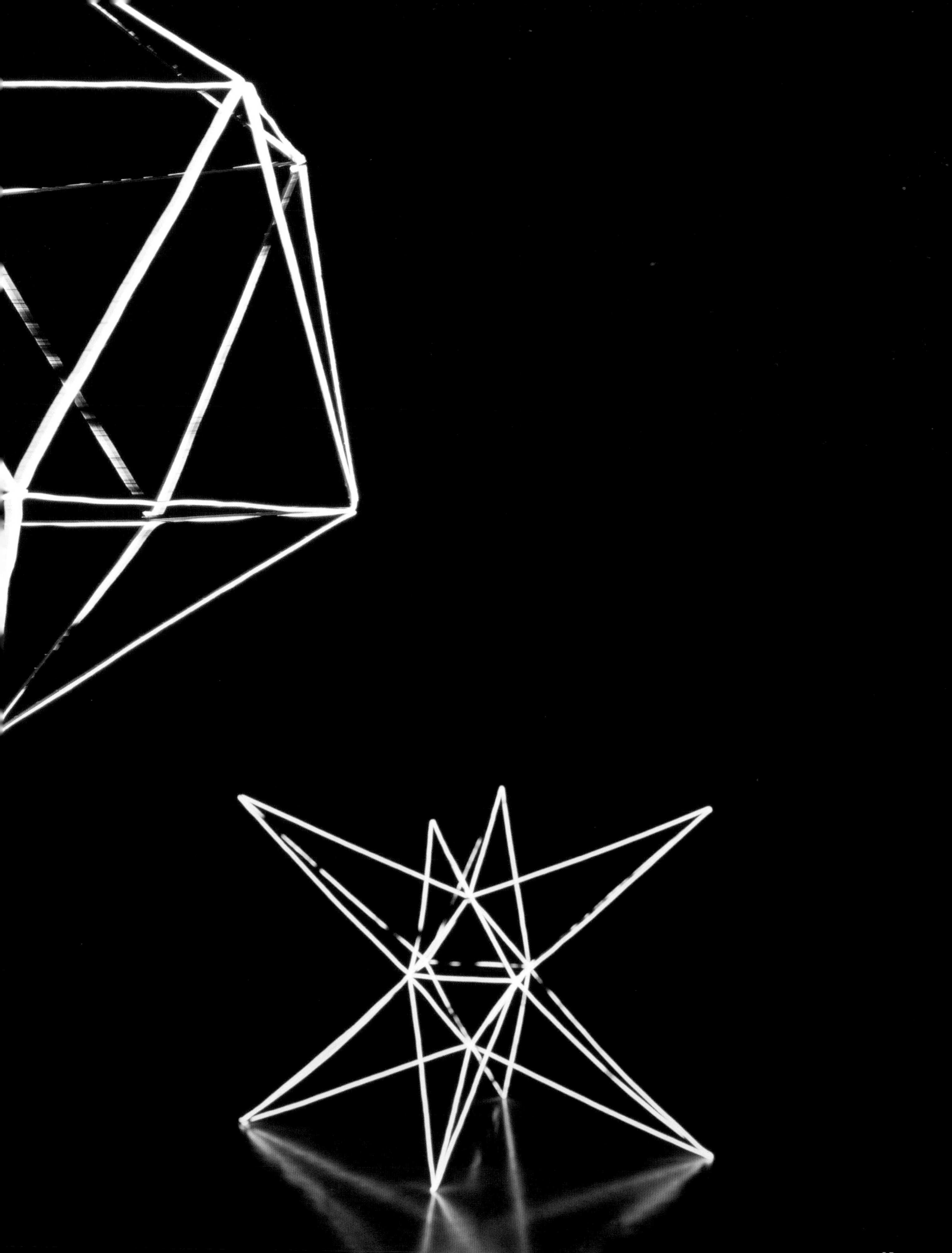

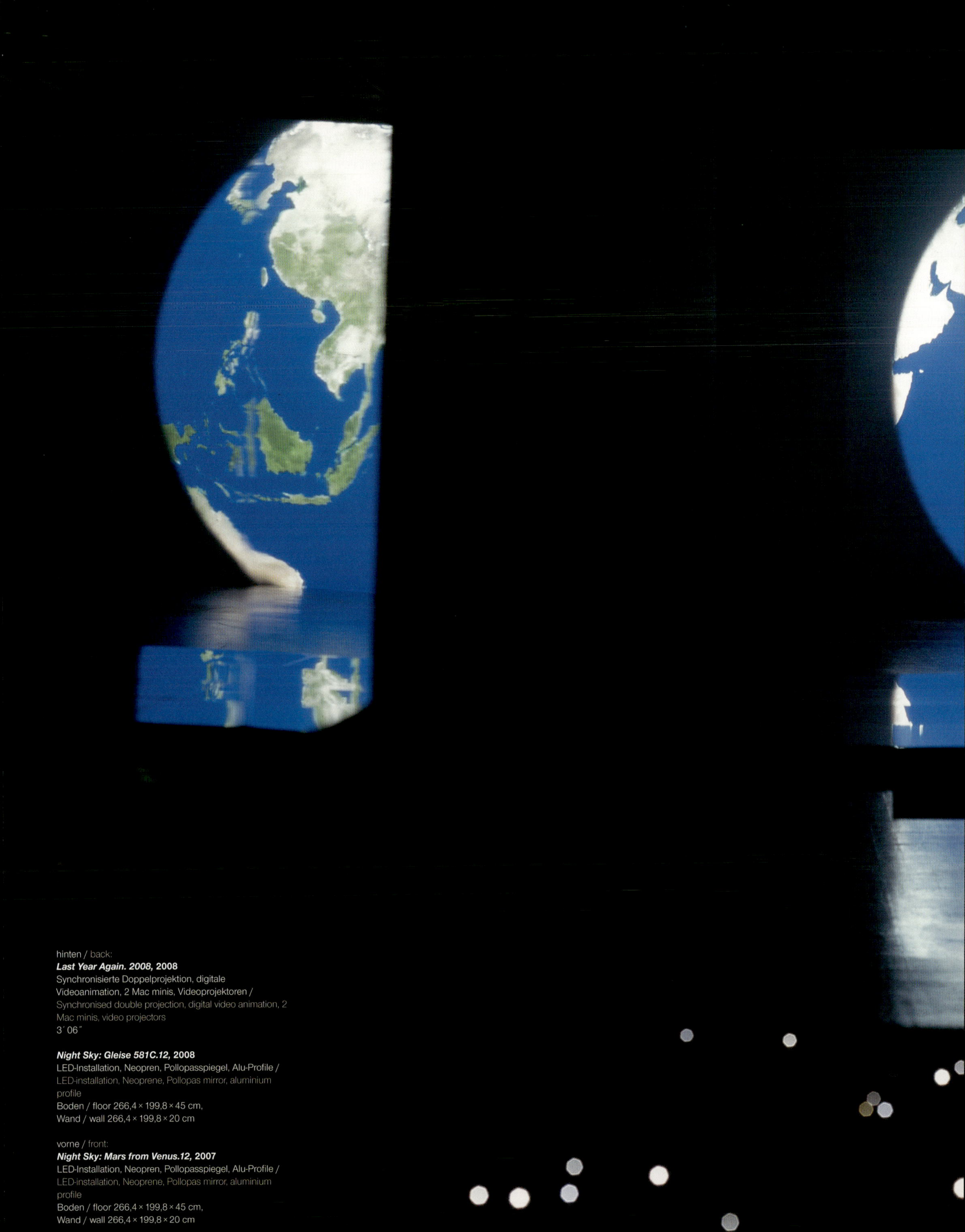

hinten / back:
***Last Year Again. 2008*, 2008**
Synchronisierte Doppelprojektion, digitale Videoanimation, 2 Mac minis, Videoprojektoren / Synchronised double projection, digital video animation, 2 Mac minis, video projectors
3´06˝

***Night Sky: Gleise 581C.12*, 2008**
LED-Installation, Neopren, Pollopasspiegel, Alu-Profile / LED-installation, Neoprene, Pollopas mirror, aluminium profile
Boden / floor 266,4 × 199,8 × 45 cm,
Wand / wall 266,4 × 199,8 × 20 cm

vorne / front:
***Night Sky: Mars from Venus.12*, 2007**
LED-Installation, Neopren, Pollopasspiegel, Alu-Profile / LED-installation, Neoprene, Pollopas mirror, aluminium profile
Boden / floor 266,4 × 199,8 × 45 cm,
Wand / wall 266,4 × 199,8 × 20 cm

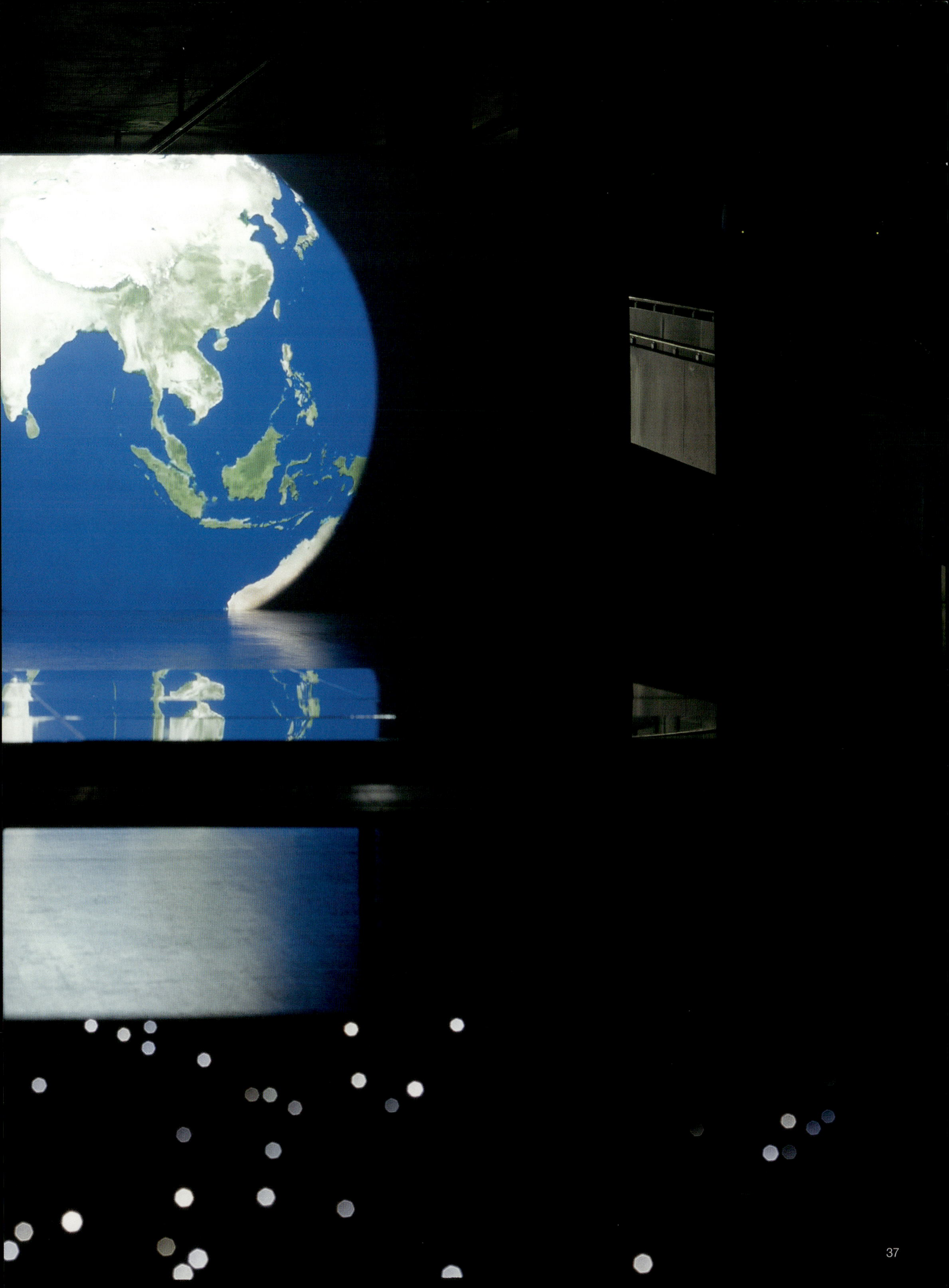

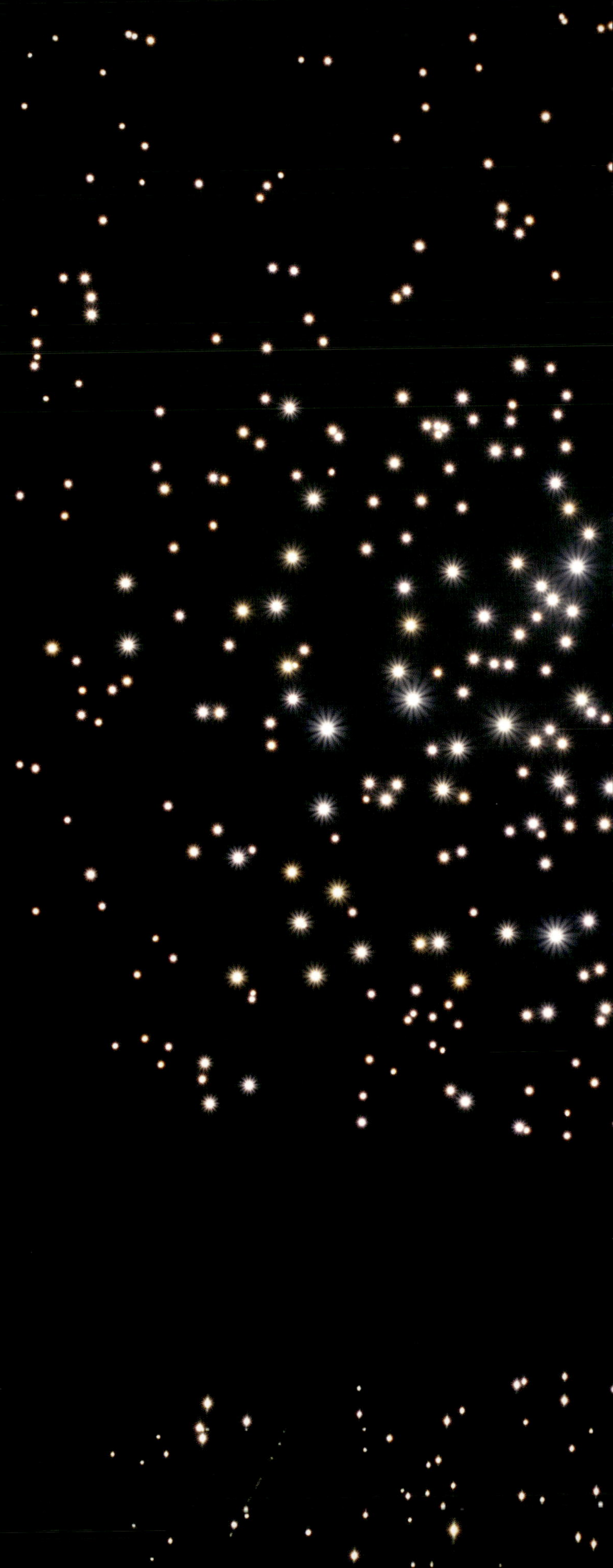

***Night Sky: E.T. from Mercury.12*, 2007**
LED-Installation, Neopren, Pollopasspiegel, Alu-Profile /
LED-installation, Neoprene, Pollopas mirror, aluminium profile
Boden / floor 266,4 × 199,8 × 45 cm,
Wand / wall 266,4 × 199,8 × 20 cm

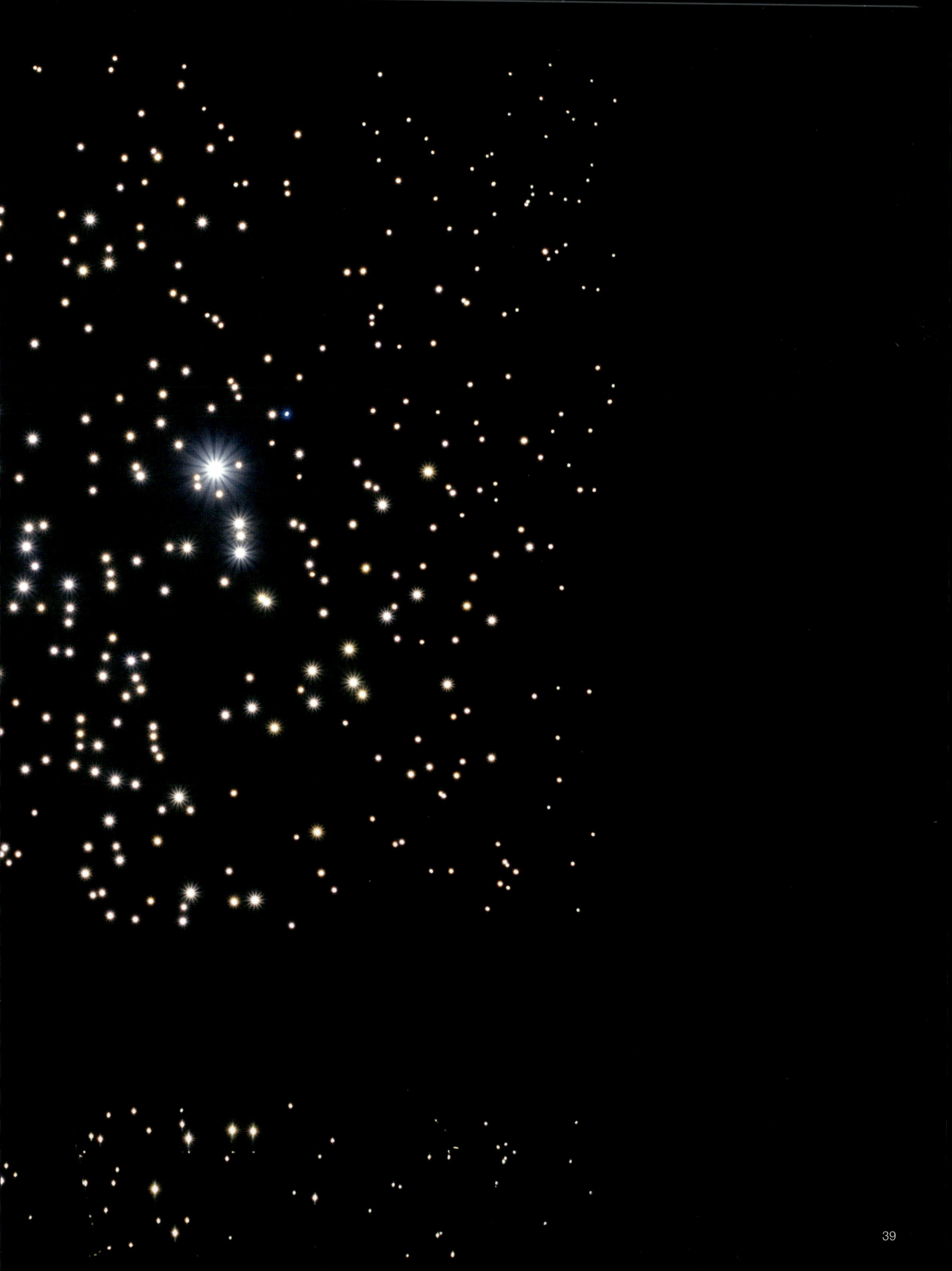

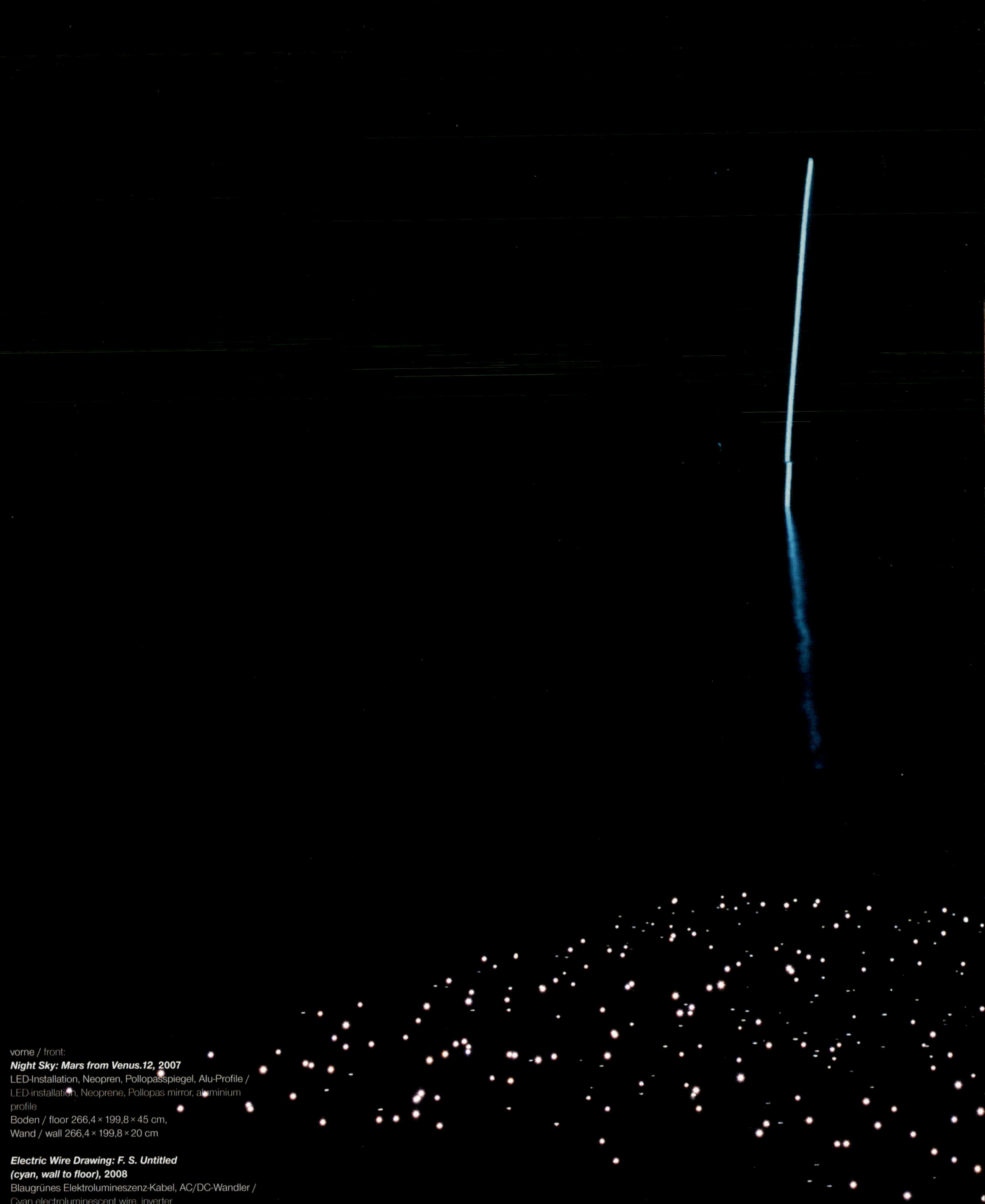

vorne / front:
***Night Sky: Mars from Venus.12*, 2007**
LED-Installation, Neopren, Pollopasspiegel, Alu-Profile / LED-installation, Neoprene, Pollopas mirror, aluminium profile
Boden / floor 266,4 × 199,8 × 45 cm,
Wand / wall 266,4 × 199,8 × 20 cm

***Electric Wire Drawing: F. S. Untitled (cyan, wall to floor)*, 2008**
Blaugrünes Elektrolumineszenz-Kabel, AC/DC-Wandler / Cyan electroluminescent wire, inverter

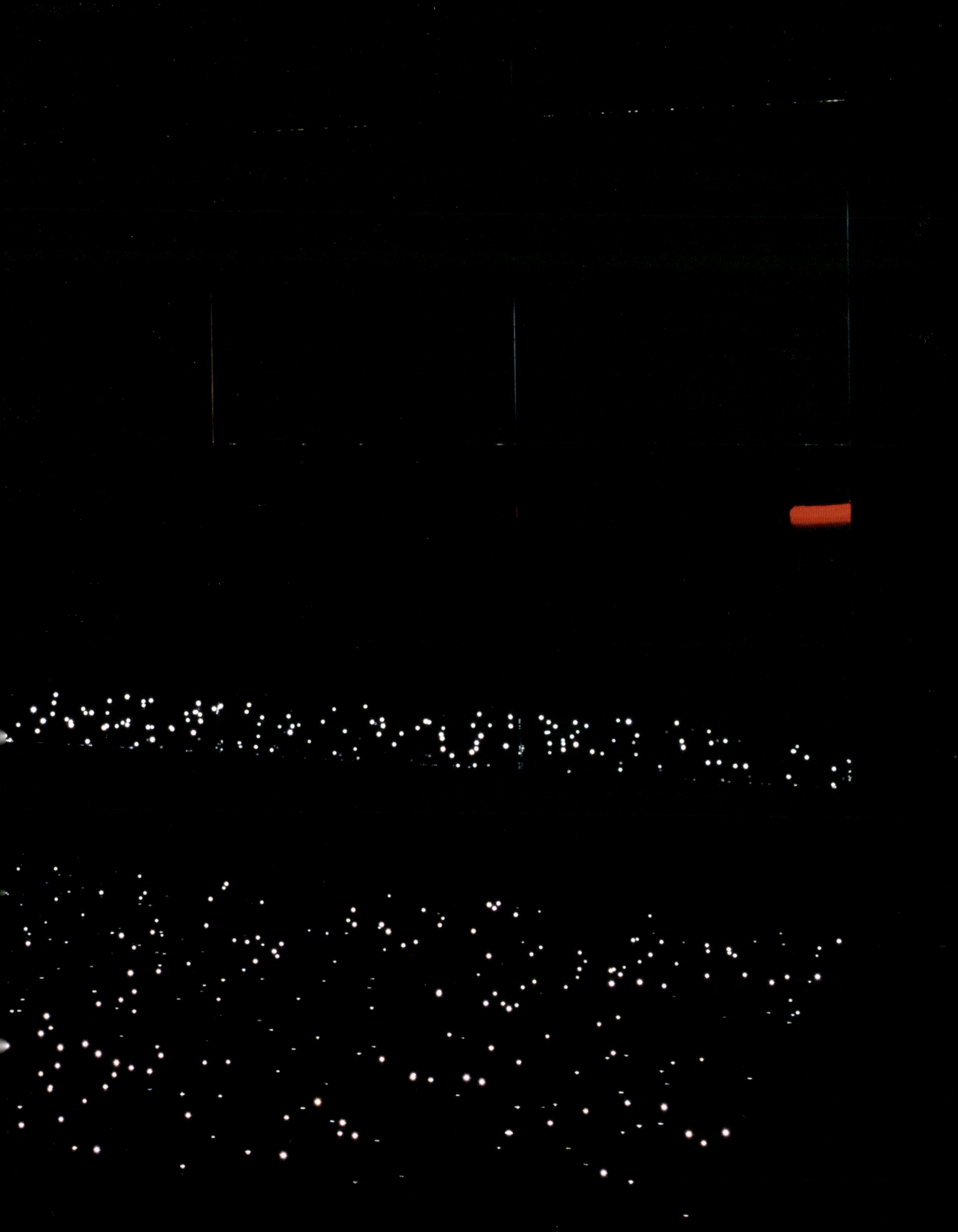

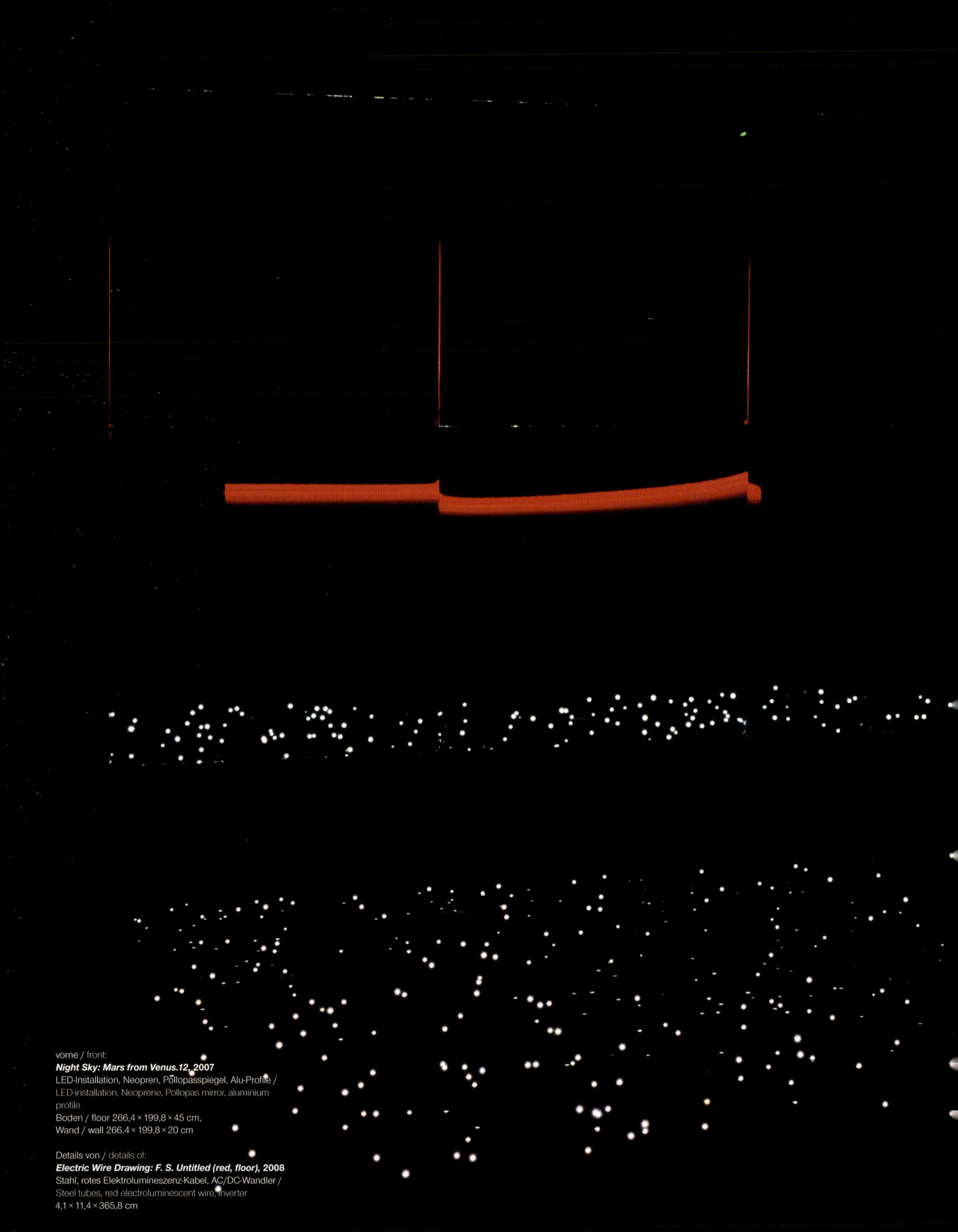

vorne / front:
***Night Sky: Mars from Venus.12*, 2007**
LED-Installation, Neopren, Pollopasspiegel, Alu-Profile / LED-installation, Neoprene, Pollopas mirror, aluminium profile
Boden / floor 266,4 × 199,8 × 45 cm,
Wand / wall 266,4 × 199,8 × 20 cm

Details von / details of:
***Electric Wire Drawing: F. S. Untitled (red, floor)*, 2008**
Stahl, rotes Elektrolumineszenz-Kabel, AC/DC-Wandler / Steel tubes, red electroluminescent wire, inverter
4,1 × 11,4 × 365,8 cm

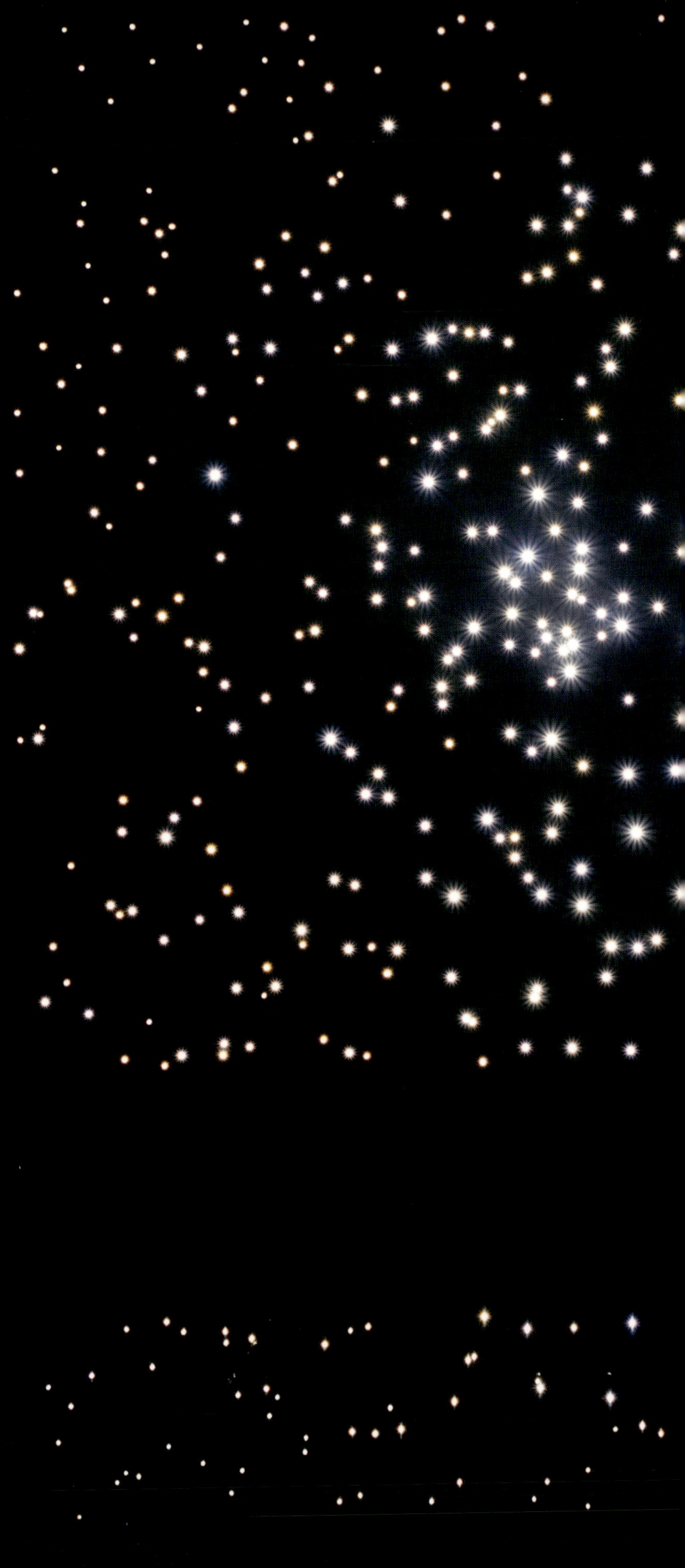

***Night Sky: Gleise 581C.12*, 2008**
LED-Installation, Neopren, Pollopasspiegel, Alu-Profile /
LED-installation, Neoprene, Pollopas mirror, aluminium profile
Boden / floor 266,4 × 199,8 × 45 cm,
Wand / wall 266,4 × 199,8 × 20 cm

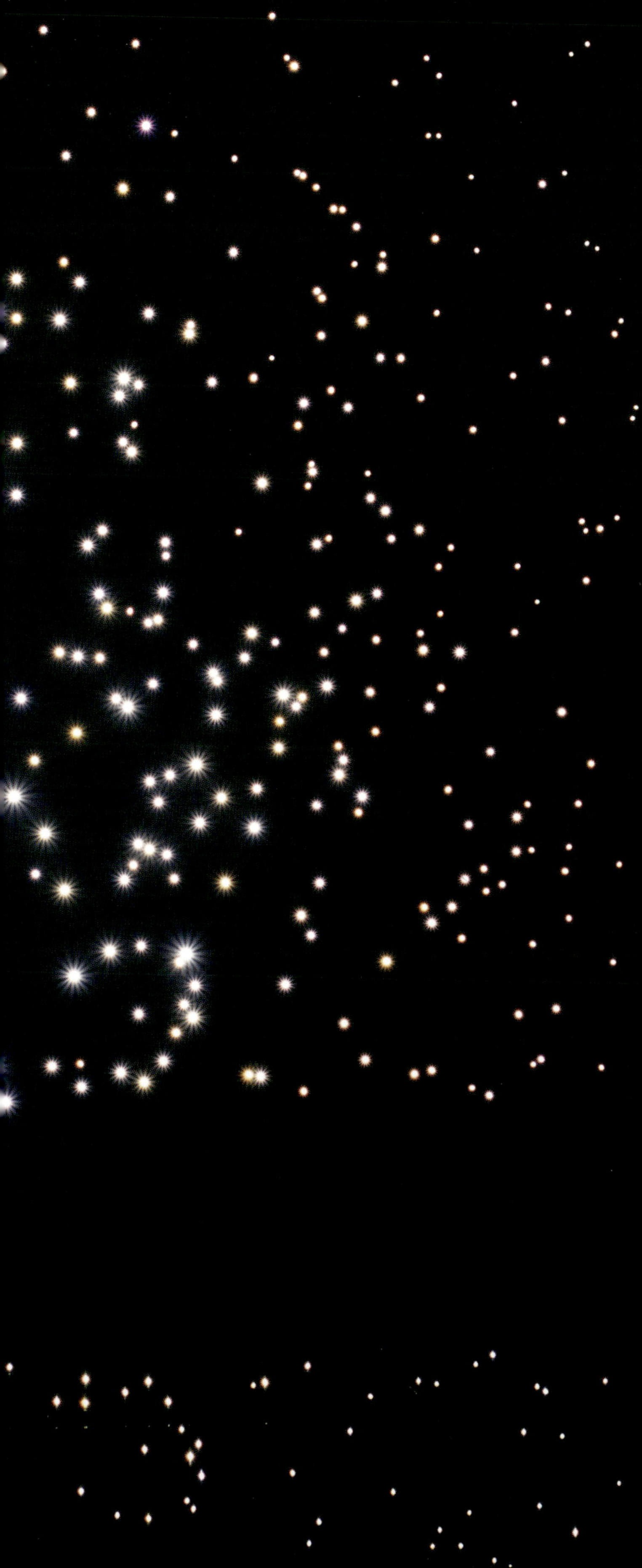

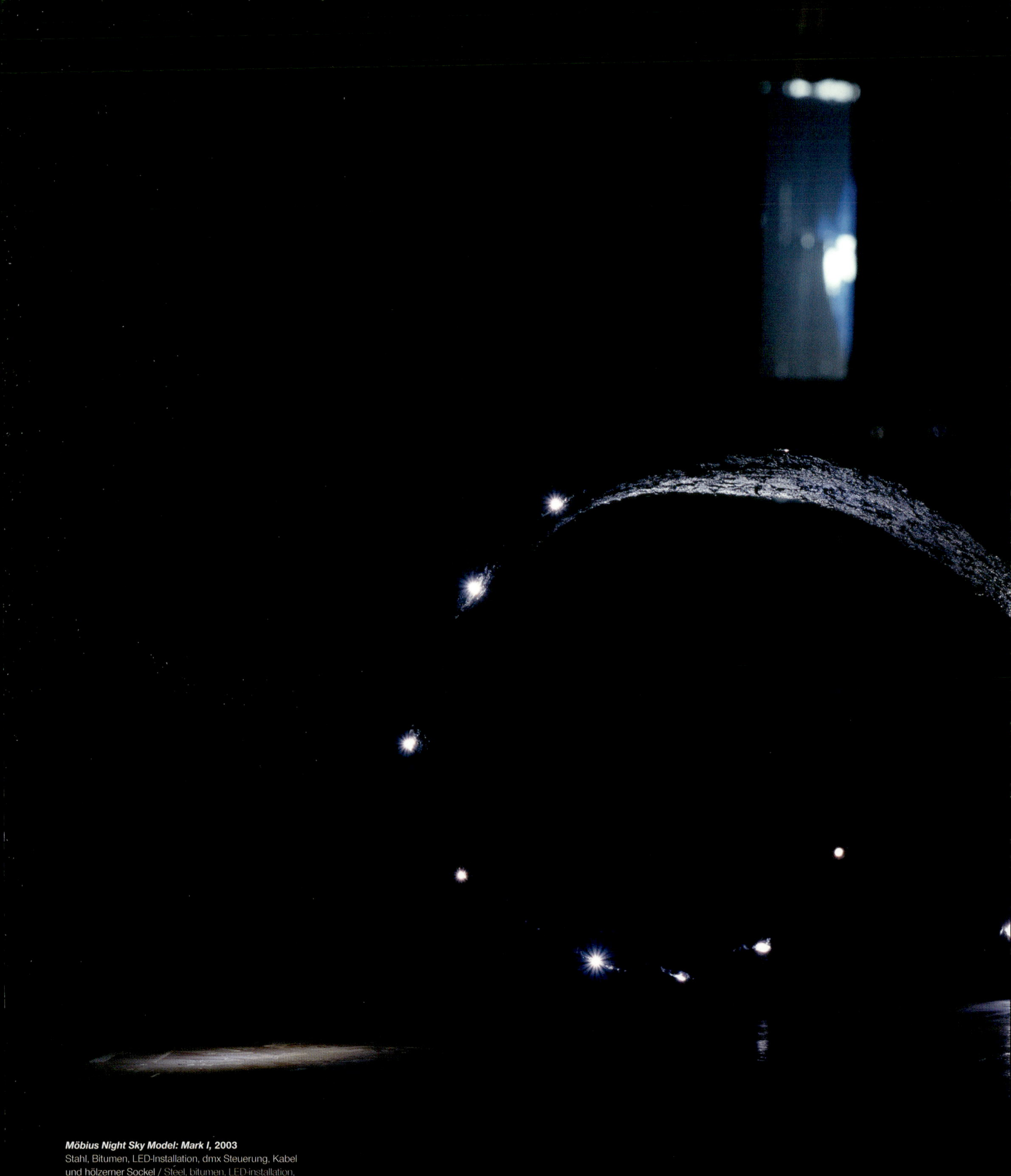

***Möbius Night Sky Model: Mark I*, 2003**
Stahl, Bitumen, LED-Installation, dmx Steuerung, Kabel und hölzerner Sockel / Steel, bitumen, LED-installation,

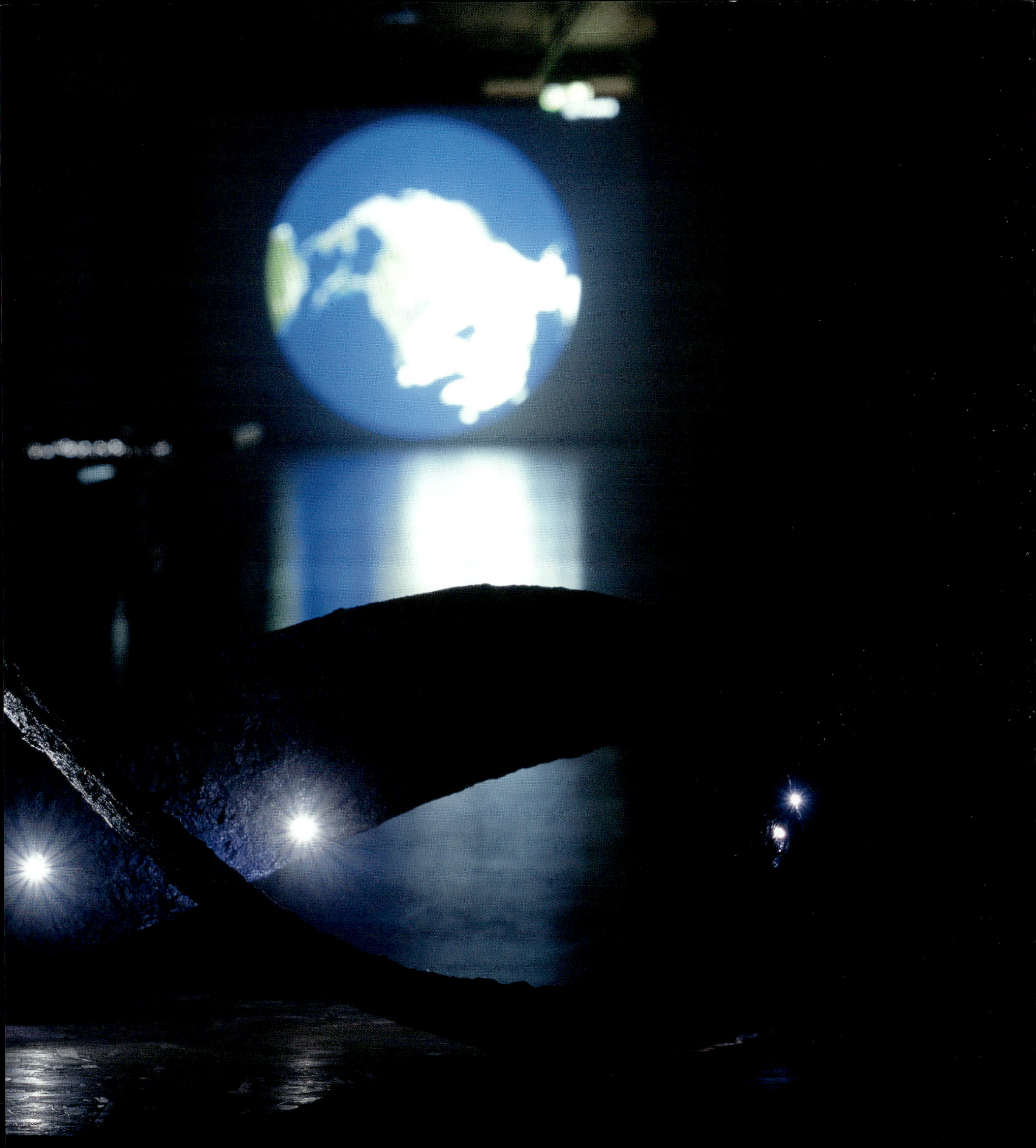

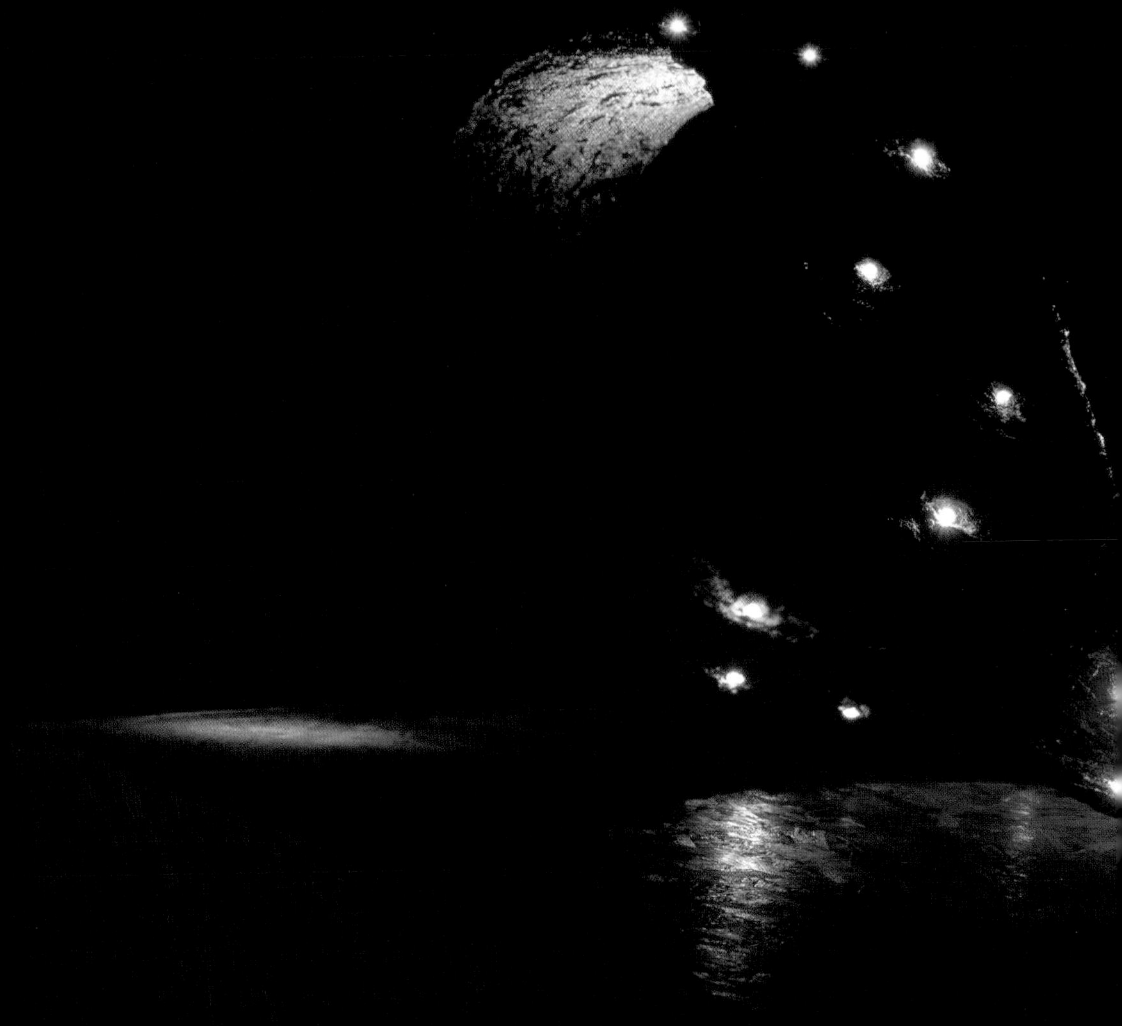

***Möbius Night Sky Model: Mark I*, 2003**
Stahl, Bitumen, LED-Installation, dmx Steuerung, Kabel und hölzerner Sockel / Steel, bitumen, LED-installation, dmx controller, cables and wooden plinth
71 × 70 × 130 cm

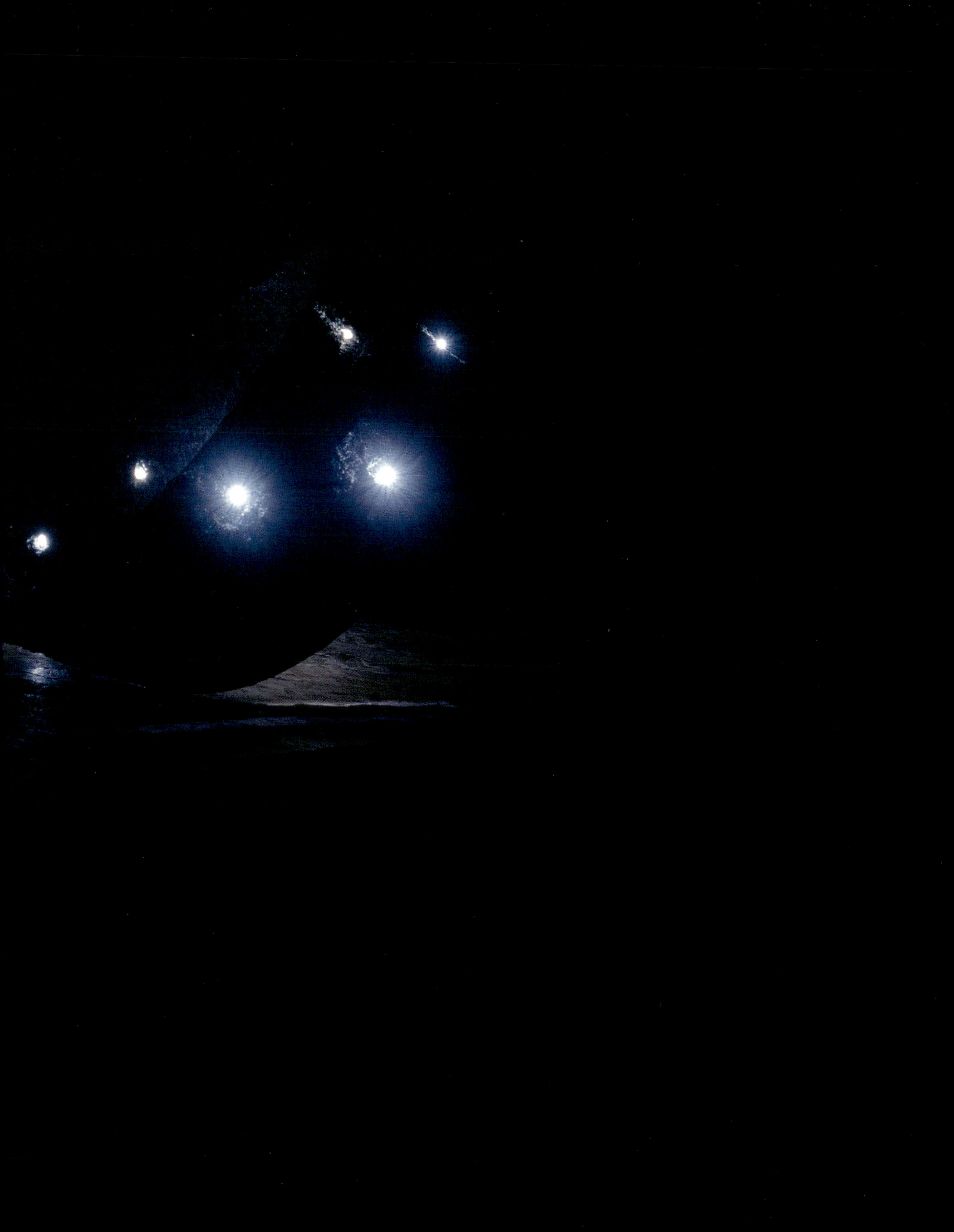

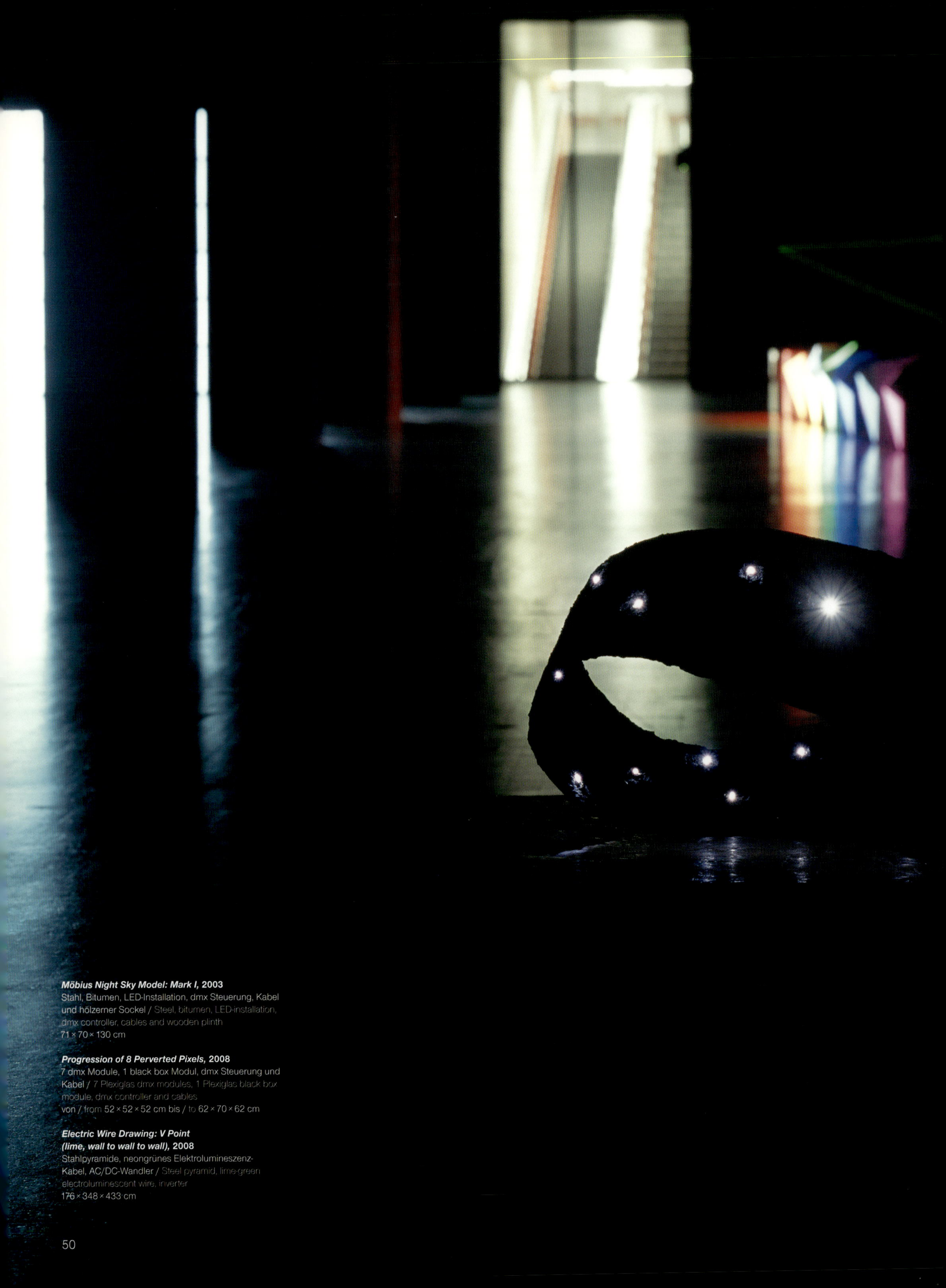

***Möbius Night Sky Model: Mark I*, 2003**
Stahl, Bitumen, LED-Installation, dmx Steuerung, Kabel und hölzerner Sockel / Steel, bitumen, LED-installation, dmx controller, cables and wooden plinth
71 × 70 × 130 cm

***Progression of 8 Perverted Pixels*, 2008**
7 dmx Module, 1 black box Modul, dmx Steuerung und Kabel / 7 Plexiglas dmx modules, 1 Plexiglas black box module, dmx controller and cables
von / from 52 × 52 × 52 cm bis / to 62 × 70 × 62 cm

***Electric Wire Drawing: V Point (lime, wall to wall to wall)*, 2008**
Stahlpyramide, neongrünes Elektrolumineszenz-Kabel, AC/DC-Wandler / Steel pyramid, lime-green electroluminescent wire, inverter
176 × 348 × 433 cm

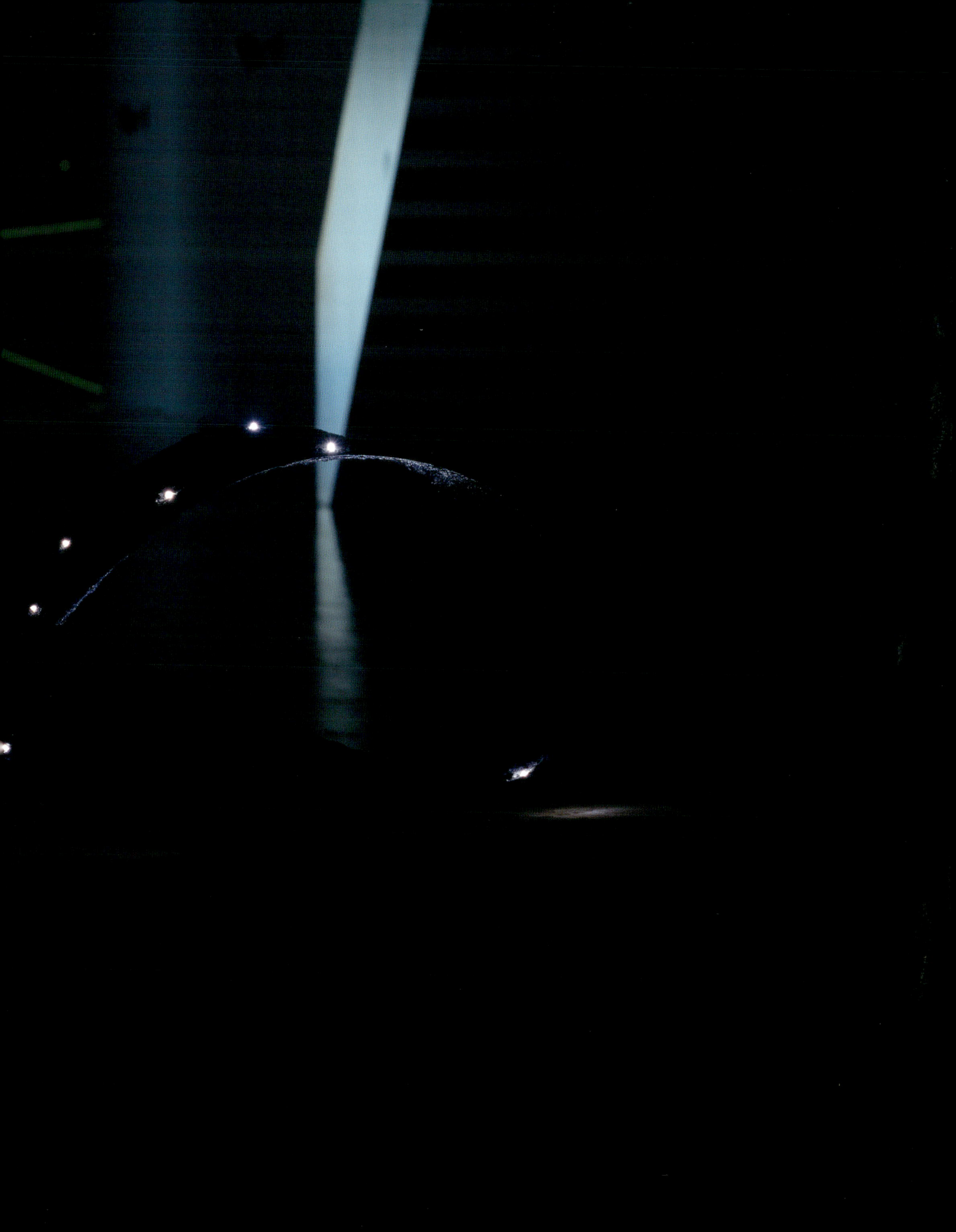

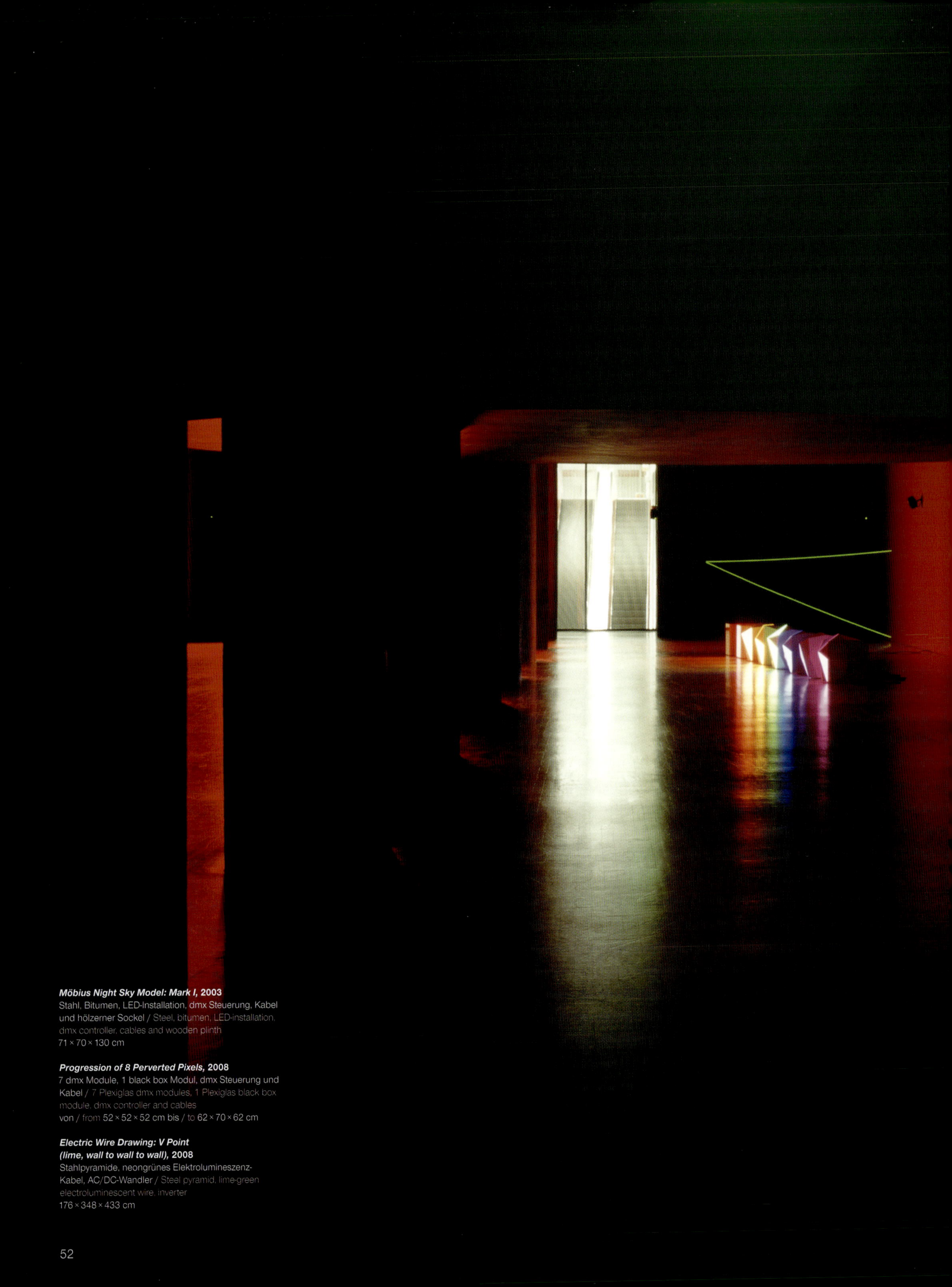

***Möbius Night Sky Model: Mark I*, 2003**
Stahl, Bitumen, LED-Installation, dmx Steuerung, Kabel und hölzerner Sockel / Steel, bitumen, LED-installation, dmx controller, cables and wooden plinth
71 × 70 × 130 cm

***Progression of 8 Perverted Pixels*, 2008**
7 dmx Module, 1 black box Modul, dmx Steuerung und Kabel / 7 Plexiglas dmx modules, 1 Plexiglas black box module, dmx controller and cables
von / from 52 × 52 × 52 cm bis / to 62 × 70 × 62 cm

***Electric Wire Drawing: V Point (lime, wall to wall to wall)*, 2008**
Stahlpyramide, neongrünes Elektrolumineszenz-Kabel, AC/DC-Wandler / Steel pyramid, lime-green electroluminescent wire, inverter
176 × 348 × 433 cm

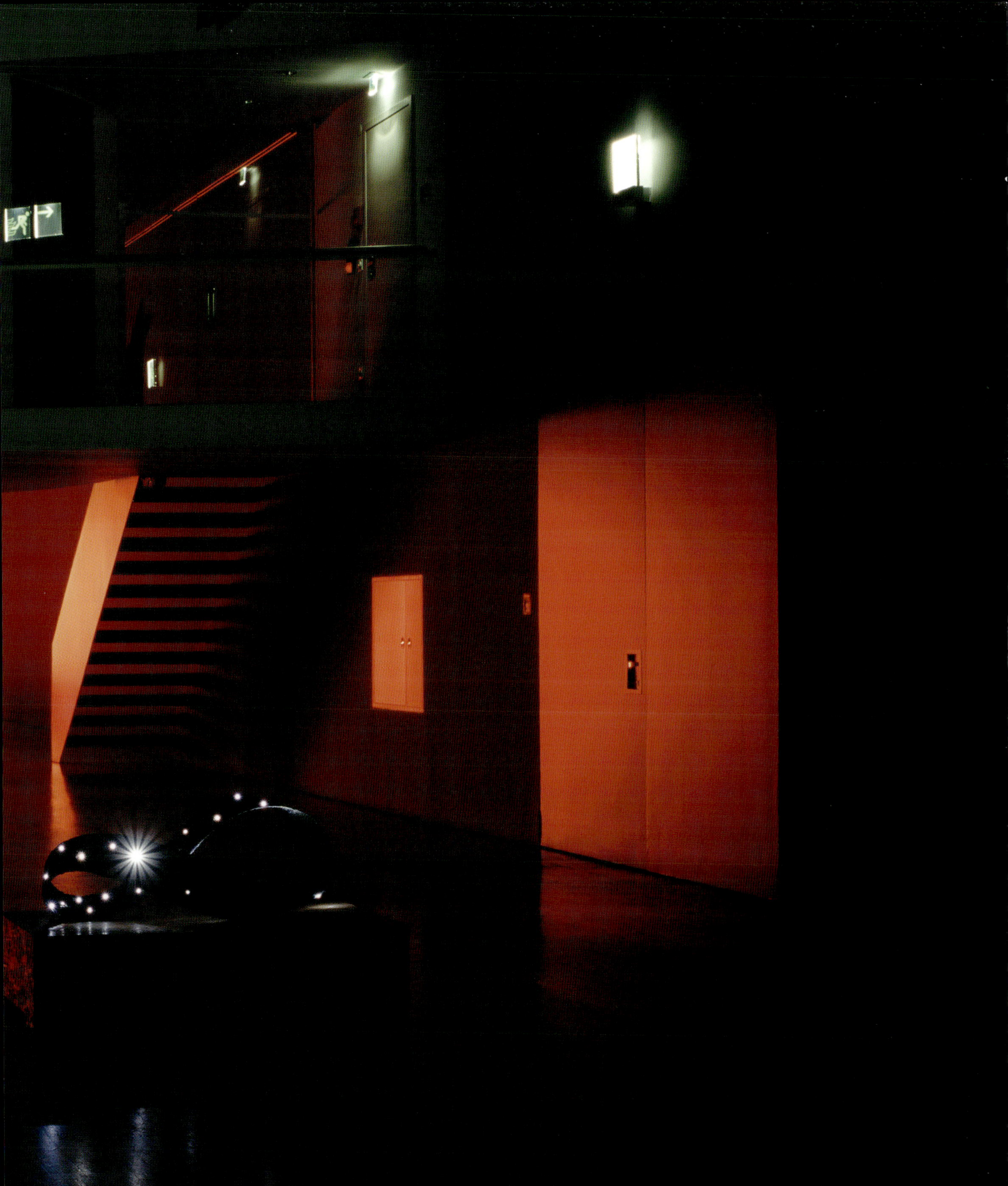

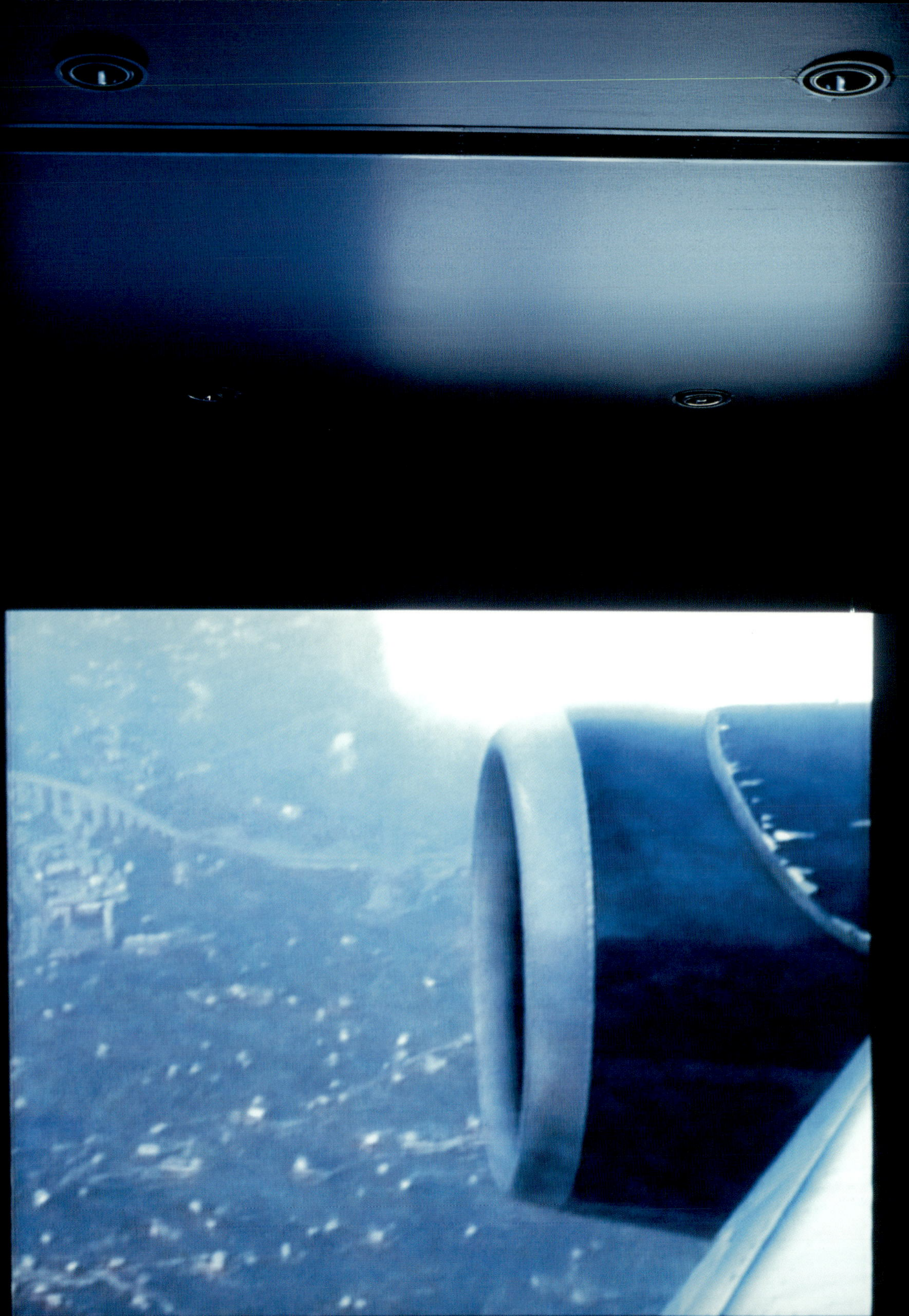

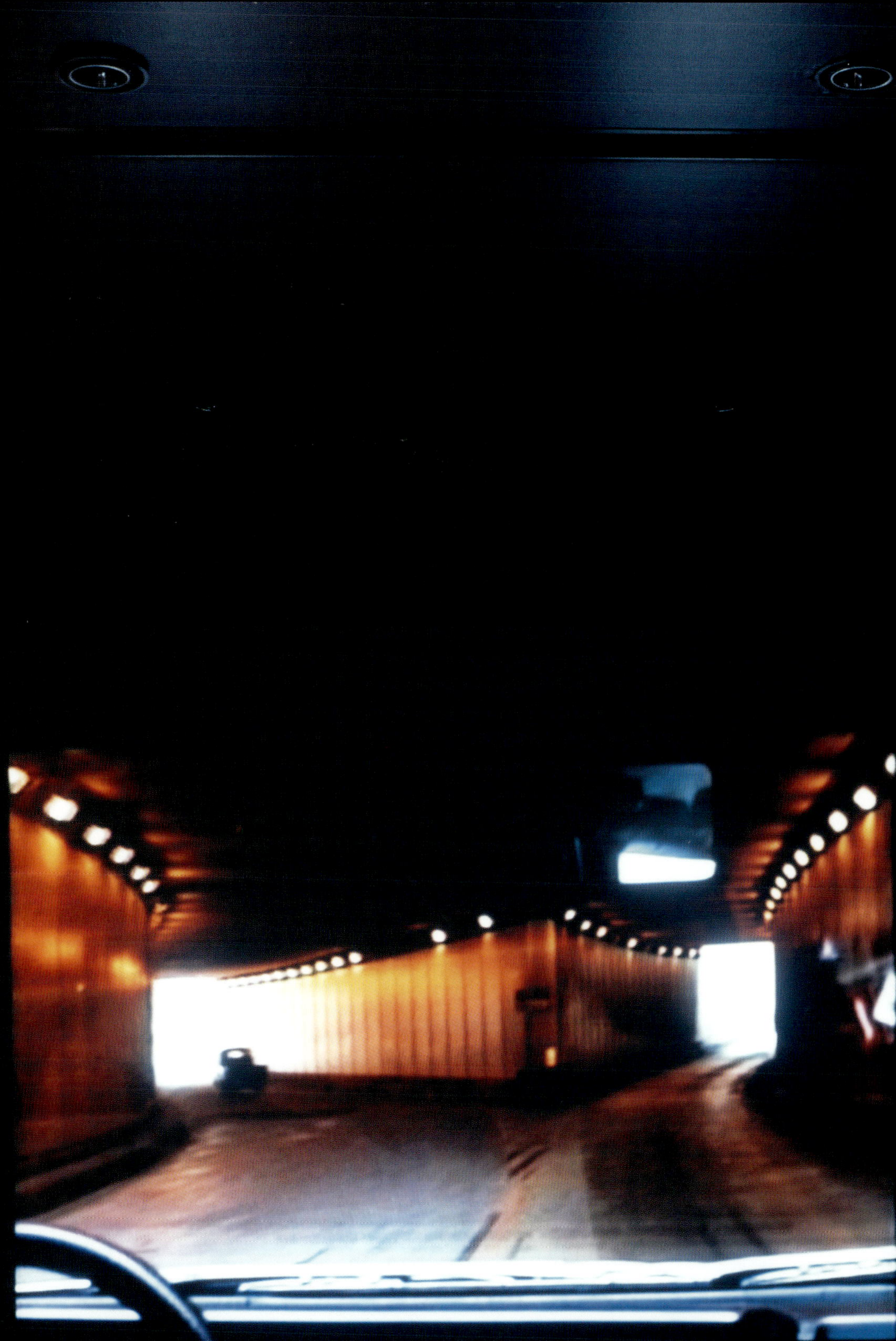

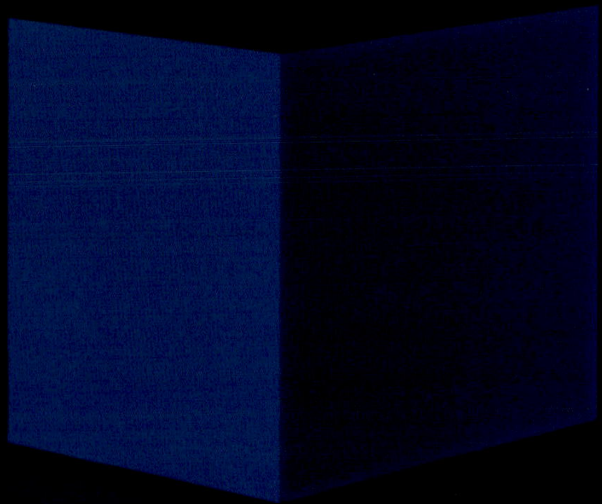

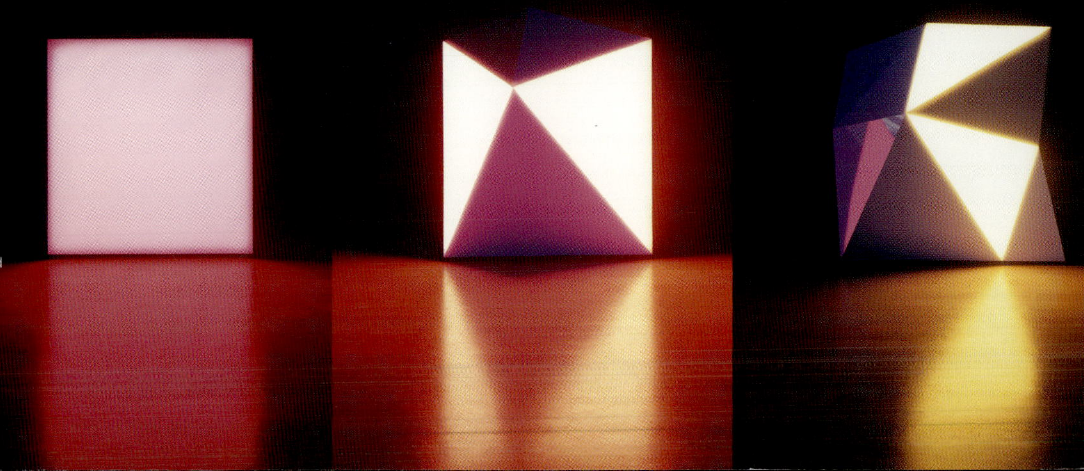

***Progression of 8 Perverted Pixels*, 2008**
7 dmx Module, 1 black box Modul, dmx Steuerung und Kabel / 7 Plexiglas dmx modules, 1 Plexiglas black box module, dmx controller and cables
von / from 52 × 52 × 52 cm bis / to 62 × 70 × 62 cm

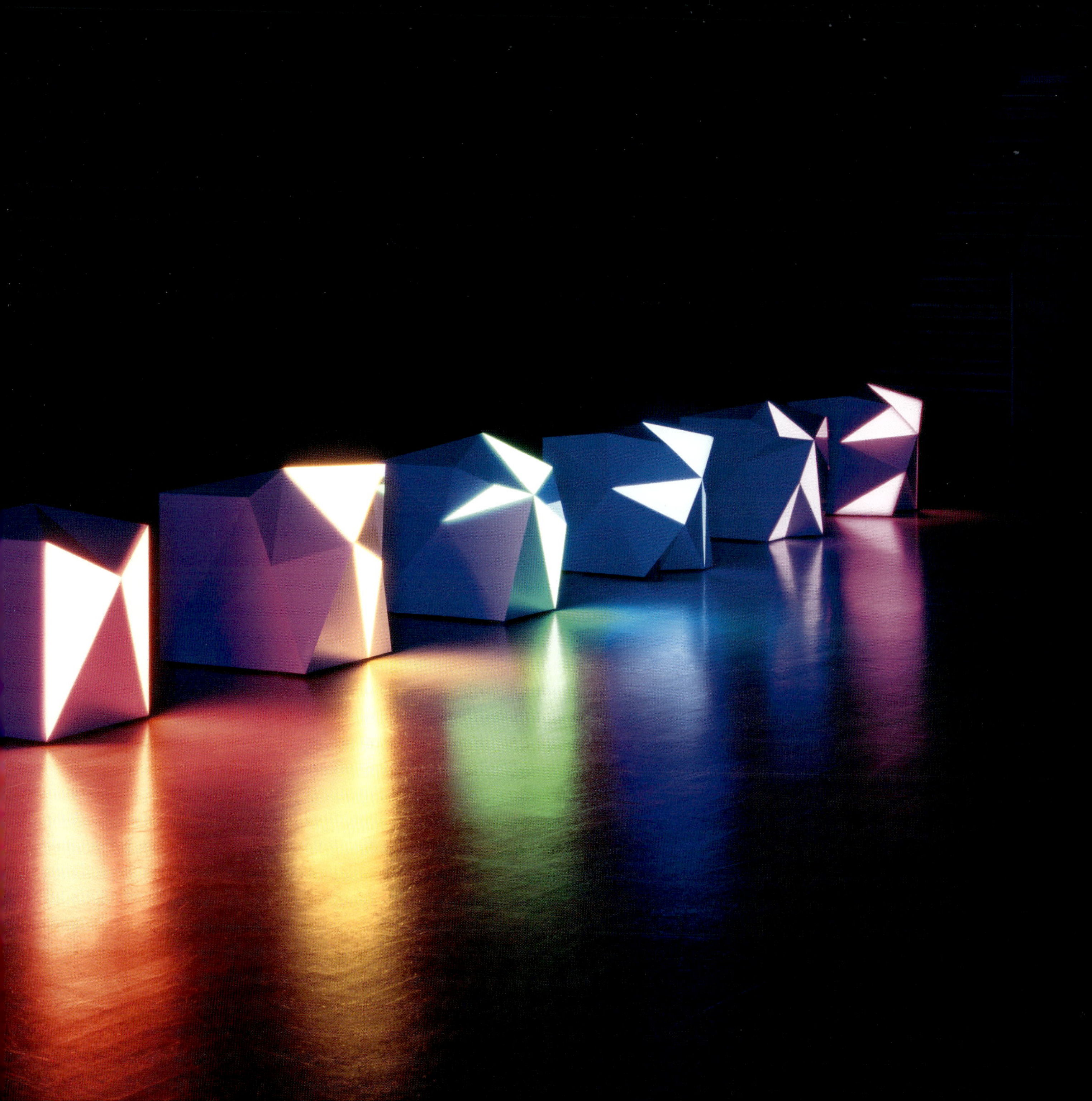

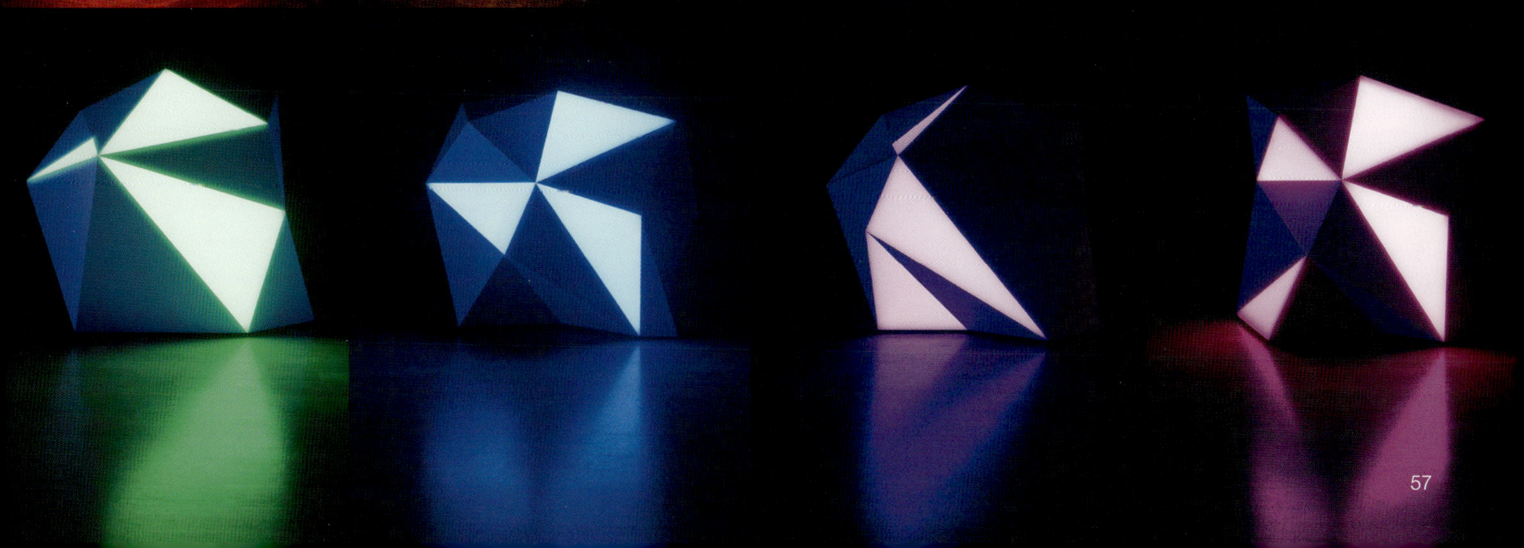

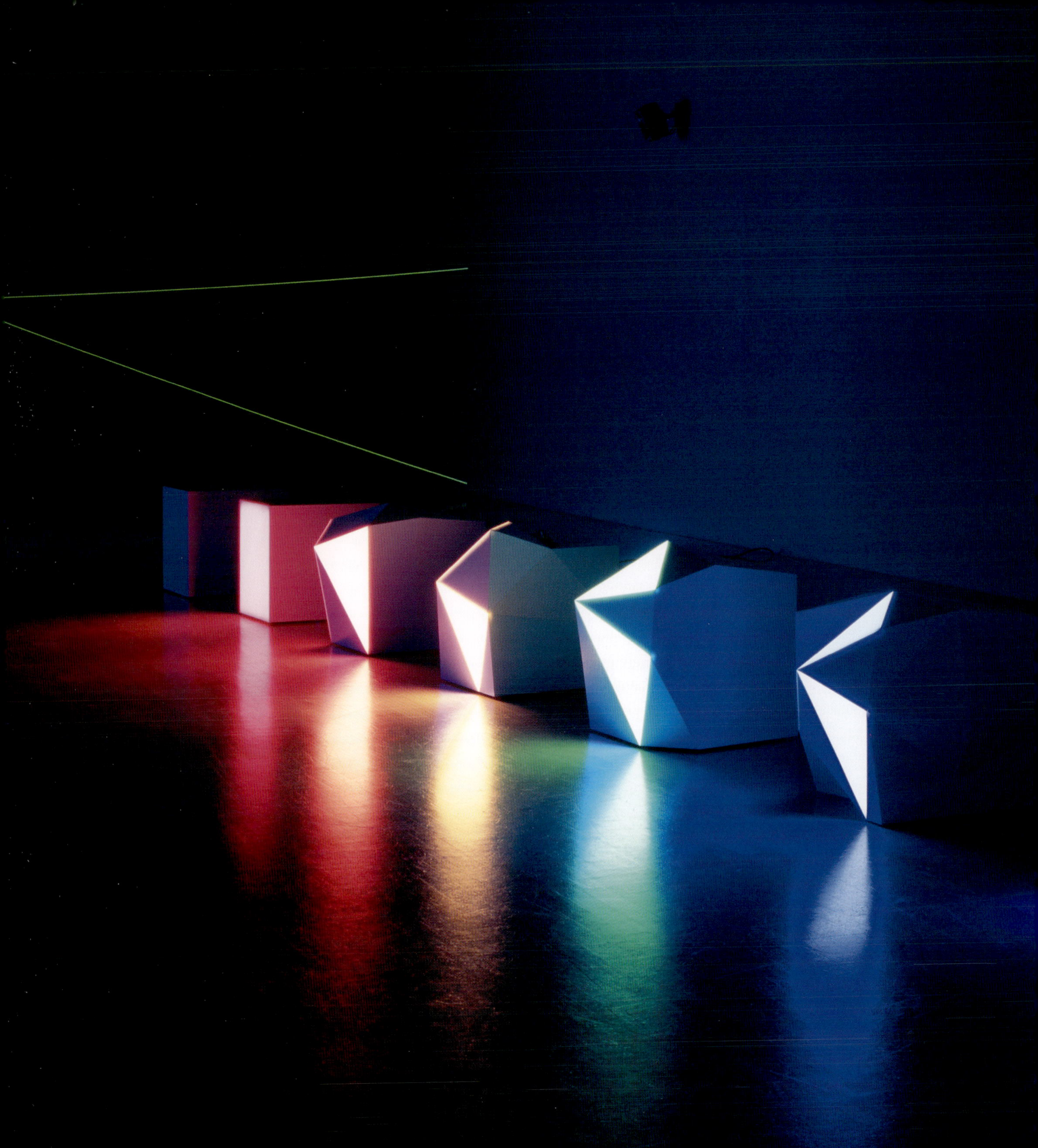

***Progression of 8 Perverted Pixels*, 2008**
7 dmx Module, 1 black box Modul, dmx Steuerung und Kabel / 7 Plexiglas dmx modules, 1 Plexiglas black box module, dmx controller and cables
von / from 52 × 52 × 52 cm bis / to 62 × 70 × 62 cm

***Electric Wire Drawing: V Point (lime, wall to wall to wall)*, 2008**
Stahlpyramide, neongrünes Elektrolumineszenz-Kabel, AC/DC-Wandler / Steel pyramid, lime-green electroluminescent wire, inverter
176 × 348 × 433 cm

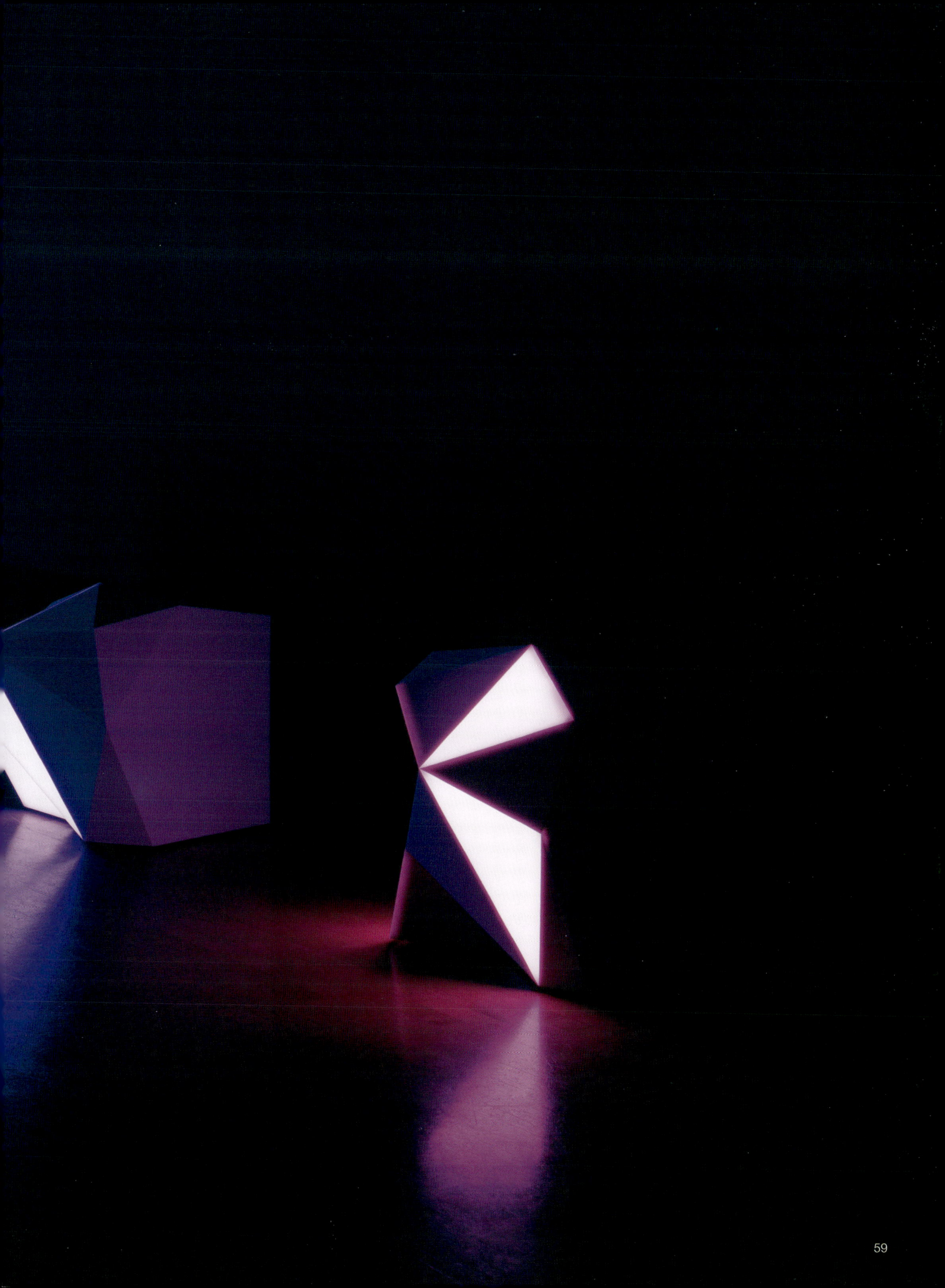

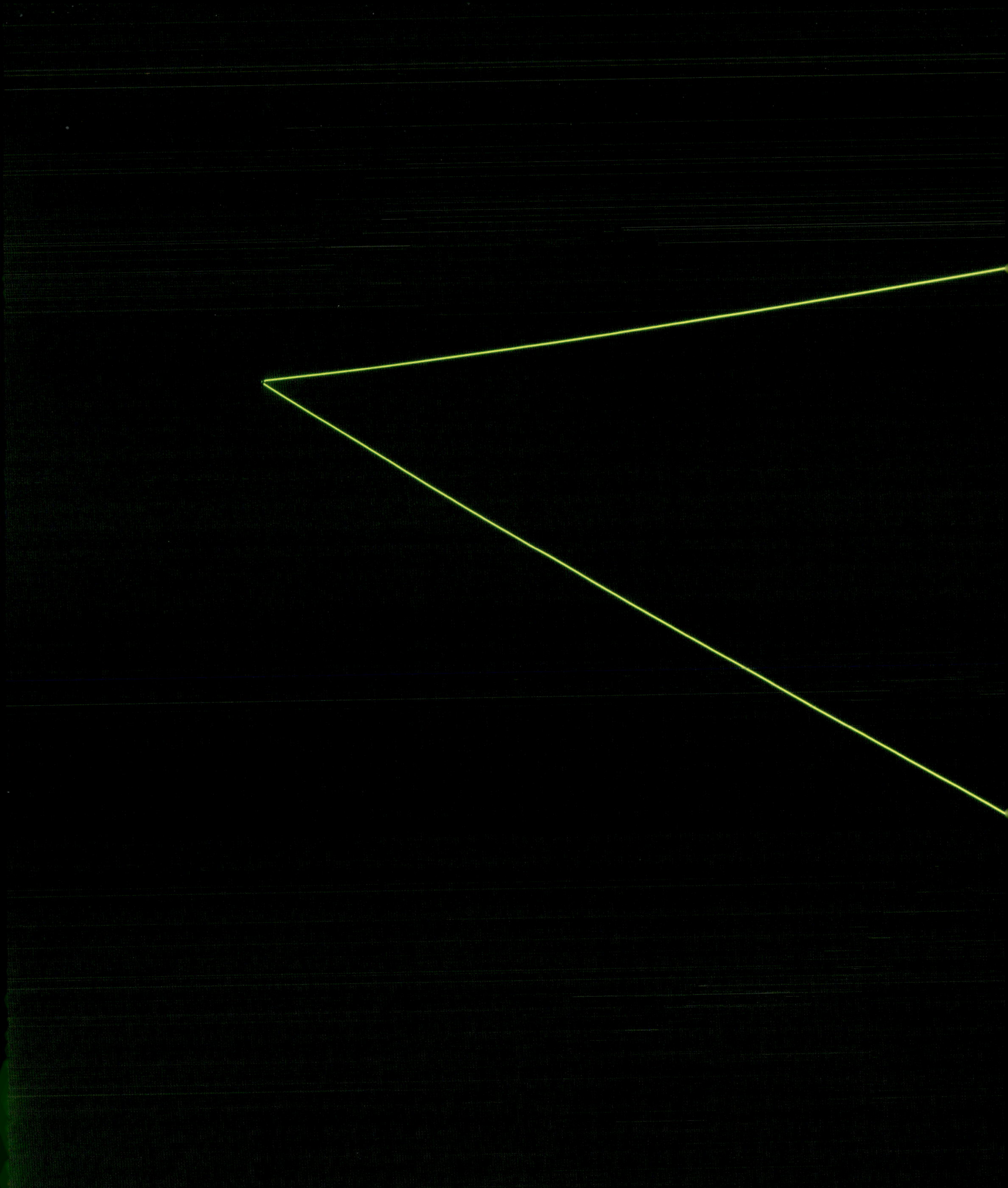

***Electric Wire Drawing: V Point (lime, wall to wall to wall)*, 2008**
Stahlpyramide, neongrünes Elektrolumineszenz-Kabel, AC/DC-Wandler / Steel pyramid, lime-green electroluminescent wire, inverter
176 × 348 × 433 cm

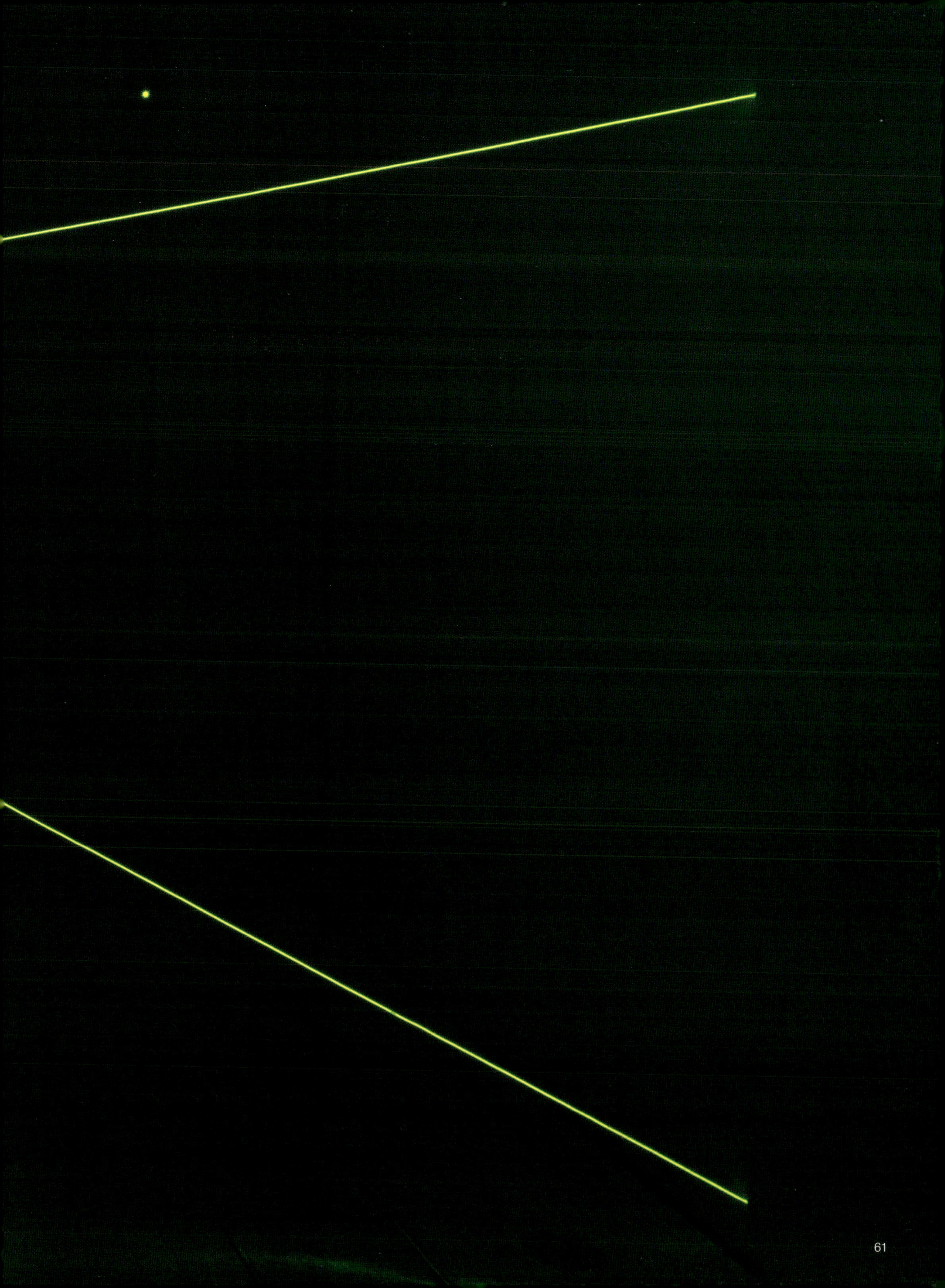

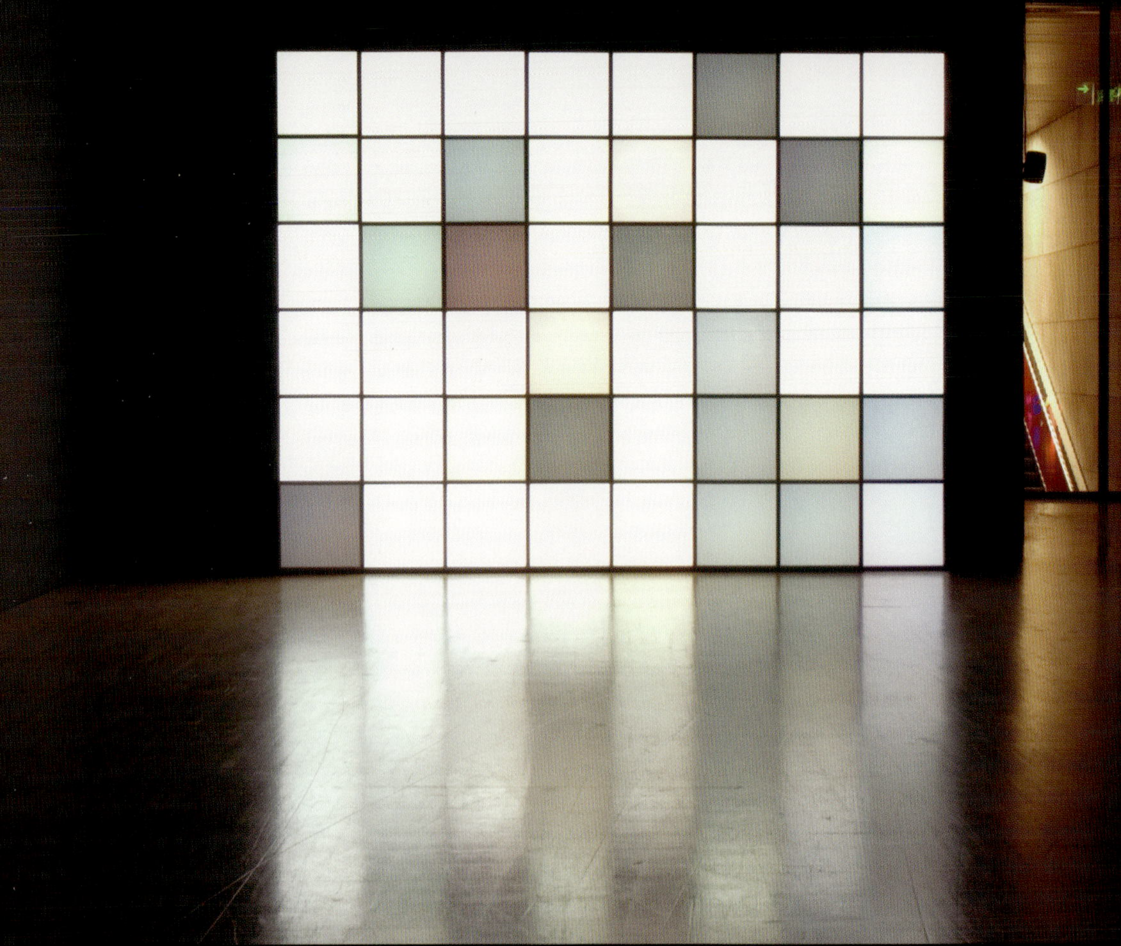

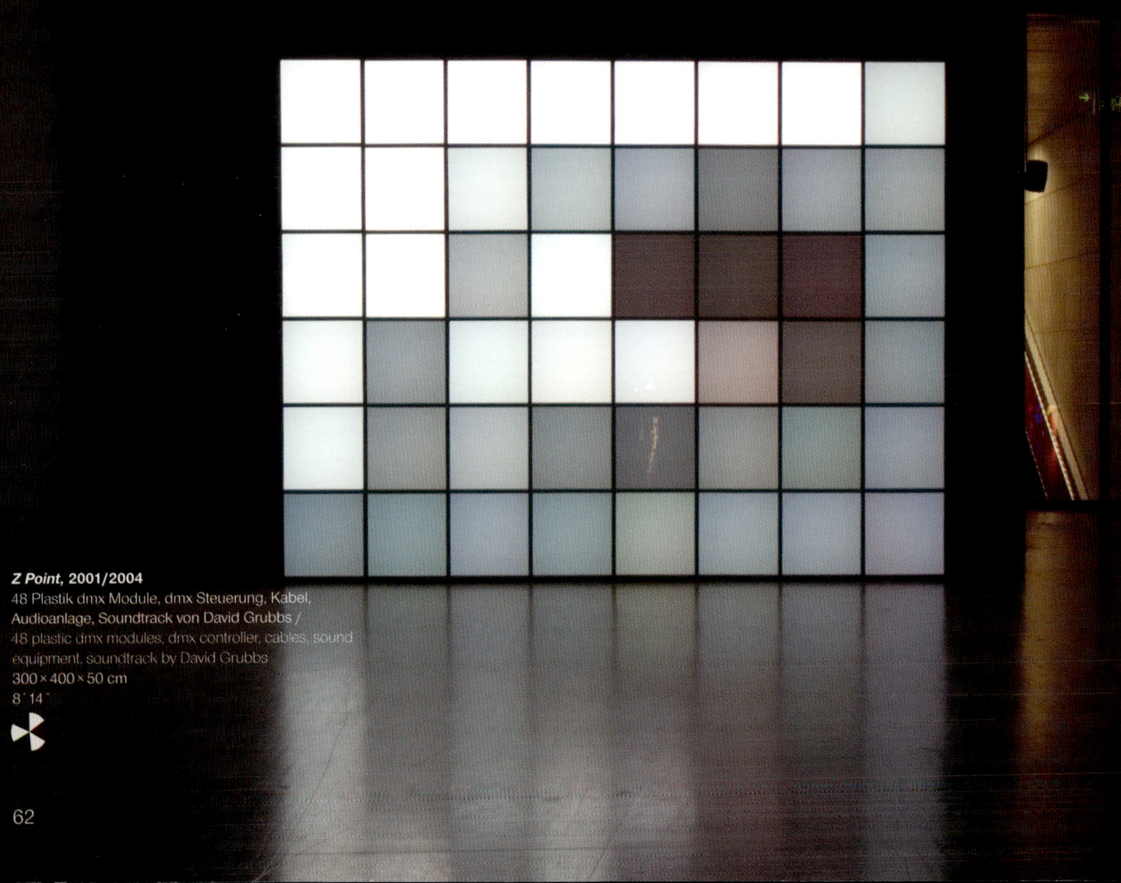

***Z Point*, 2001/2004**
48 Plastik dmx Module, dmx Steuerung, Kabel, Audioanlage, Soundtrack von David Grubbs /
48 plastic dmx modules, dmx controller, cables, sound equipment, soundtrack by David Grubbs
300 × 400 × 50 cm
8´14´´

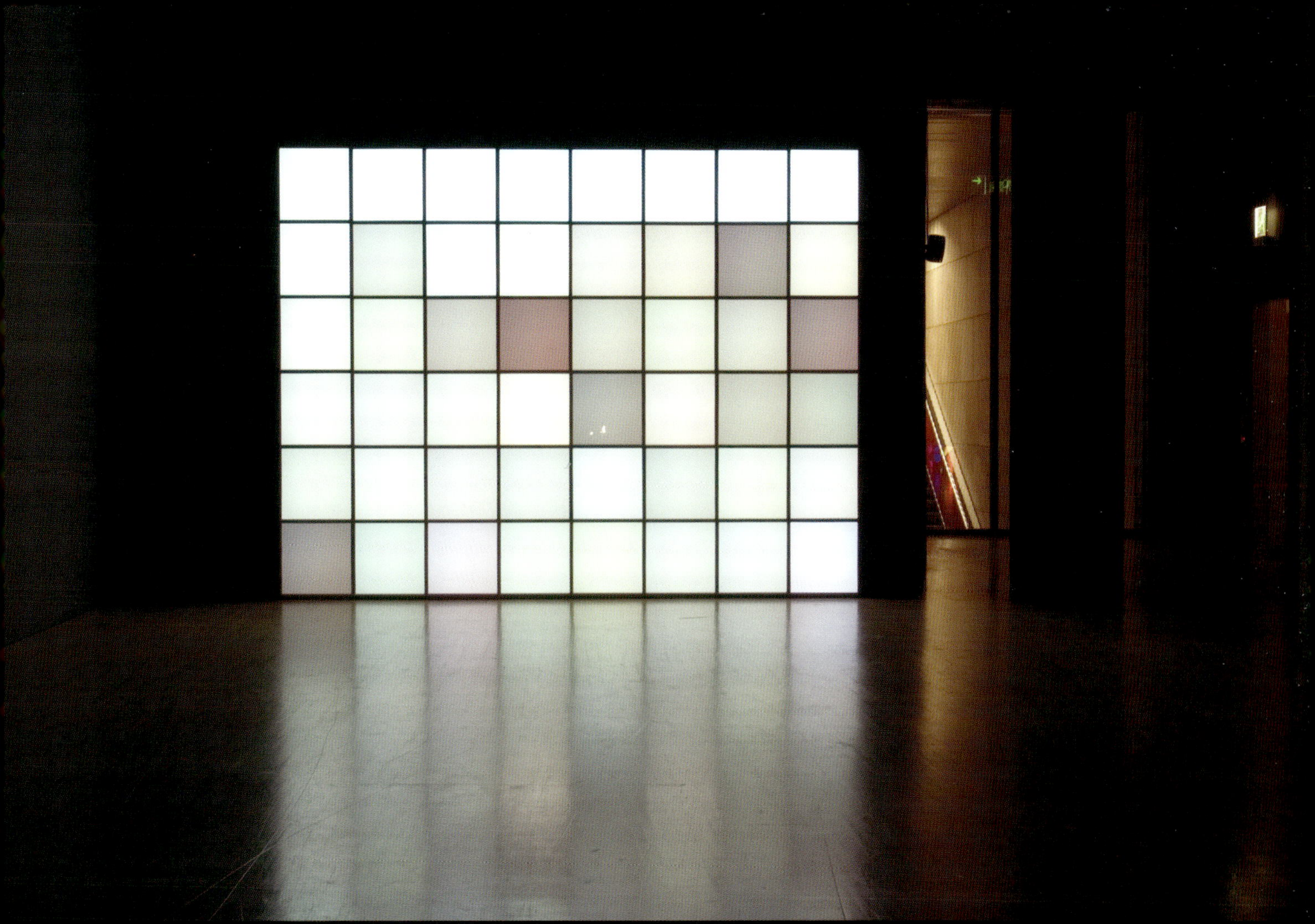

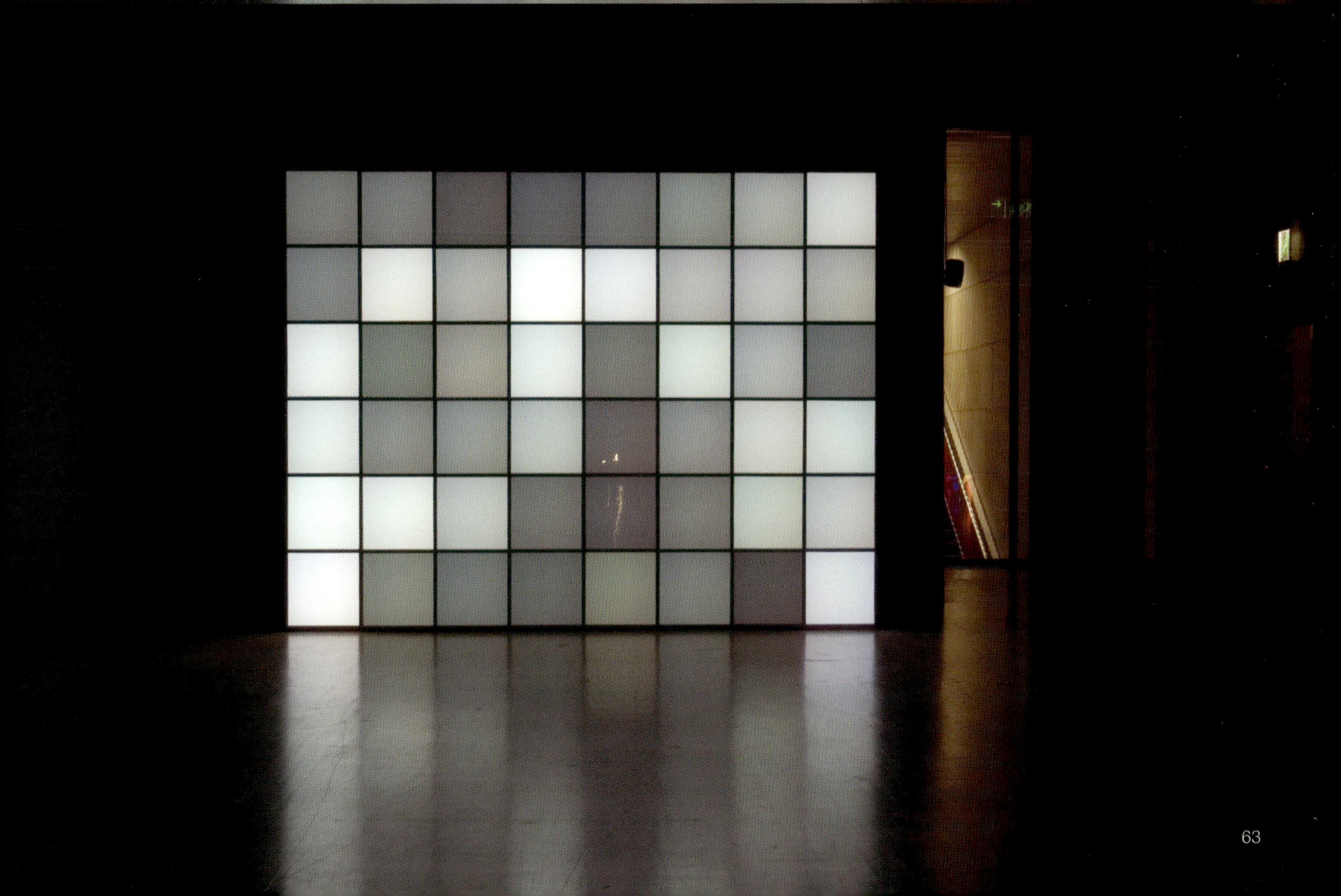

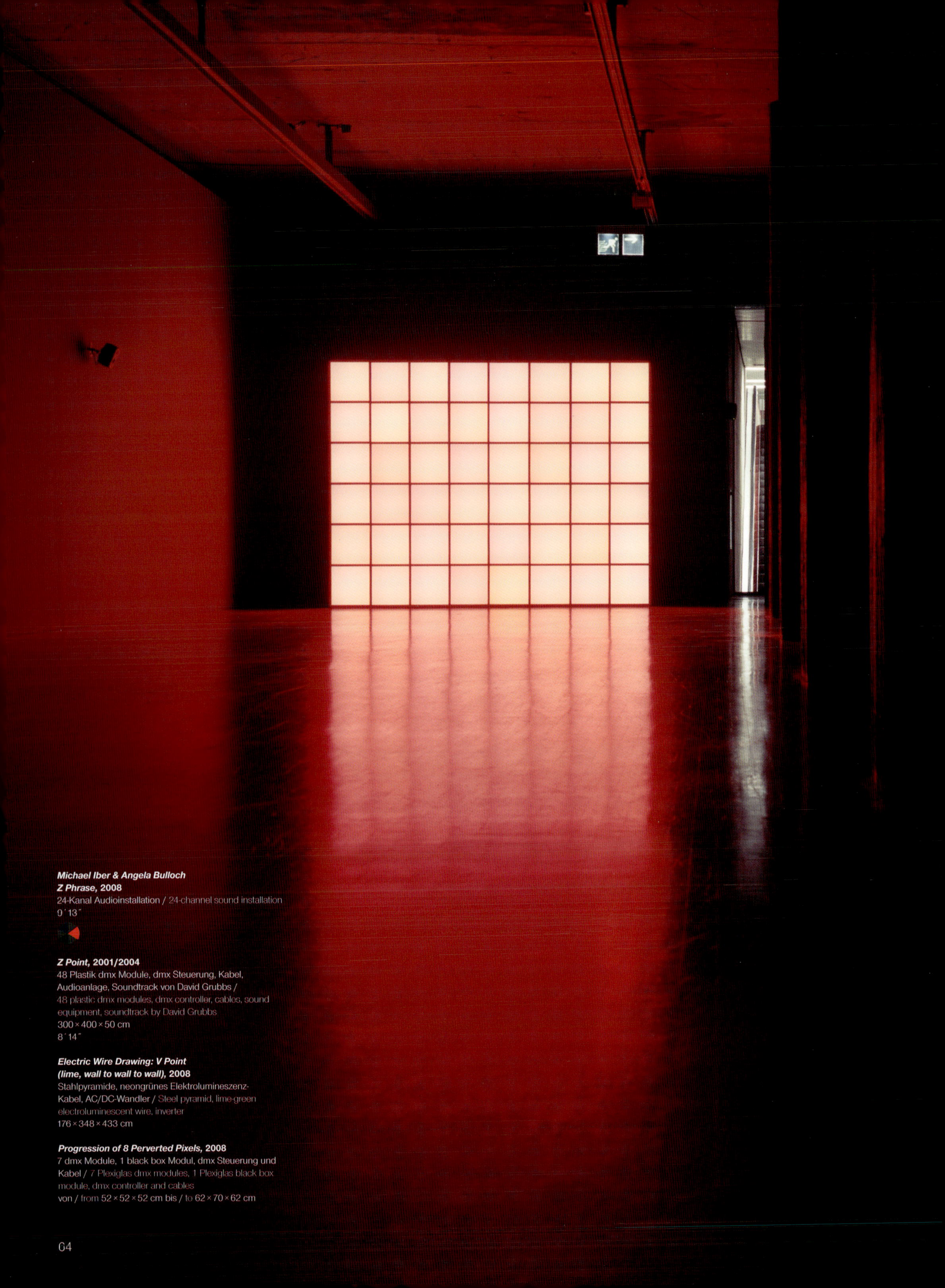

Michael Iber & Angela Bulloch
***Z Phrase*, 2008**
24-Kanal Audioinstallation / 24-channel sound installation
9´13˝

***Z Point*, 2001/2004**
48 Plastik dmx Module, dmx Steuerung, Kabel, Audioanlage, Soundtrack von David Grubbs /
48 plastic dmx modules, dmx controller, cables, sound equipment, soundtrack by David Grubbs
300 × 400 × 50 cm
8´14˝

***Electric Wire Drawing: V Point (lime, wall to wall to wall)*, 2008**
Stahlpyramide, neongrünes Elektrolumineszenz-Kabel, AC/DC-Wandler / Steel pyramid, lime-green electroluminescent wire, inverter
176 × 348 × 433 cm

***Progression of 8 Perverted Pixels*, 2008**
7 dmx Module, 1 black box Modul, dmx Steuerung und Kabel / 7 Plexiglas dmx modules, 1 Plexiglas black box module, dmx controller and cables
von / from 52 × 52 × 52 cm bis / to 62 × 70 × 62 cm

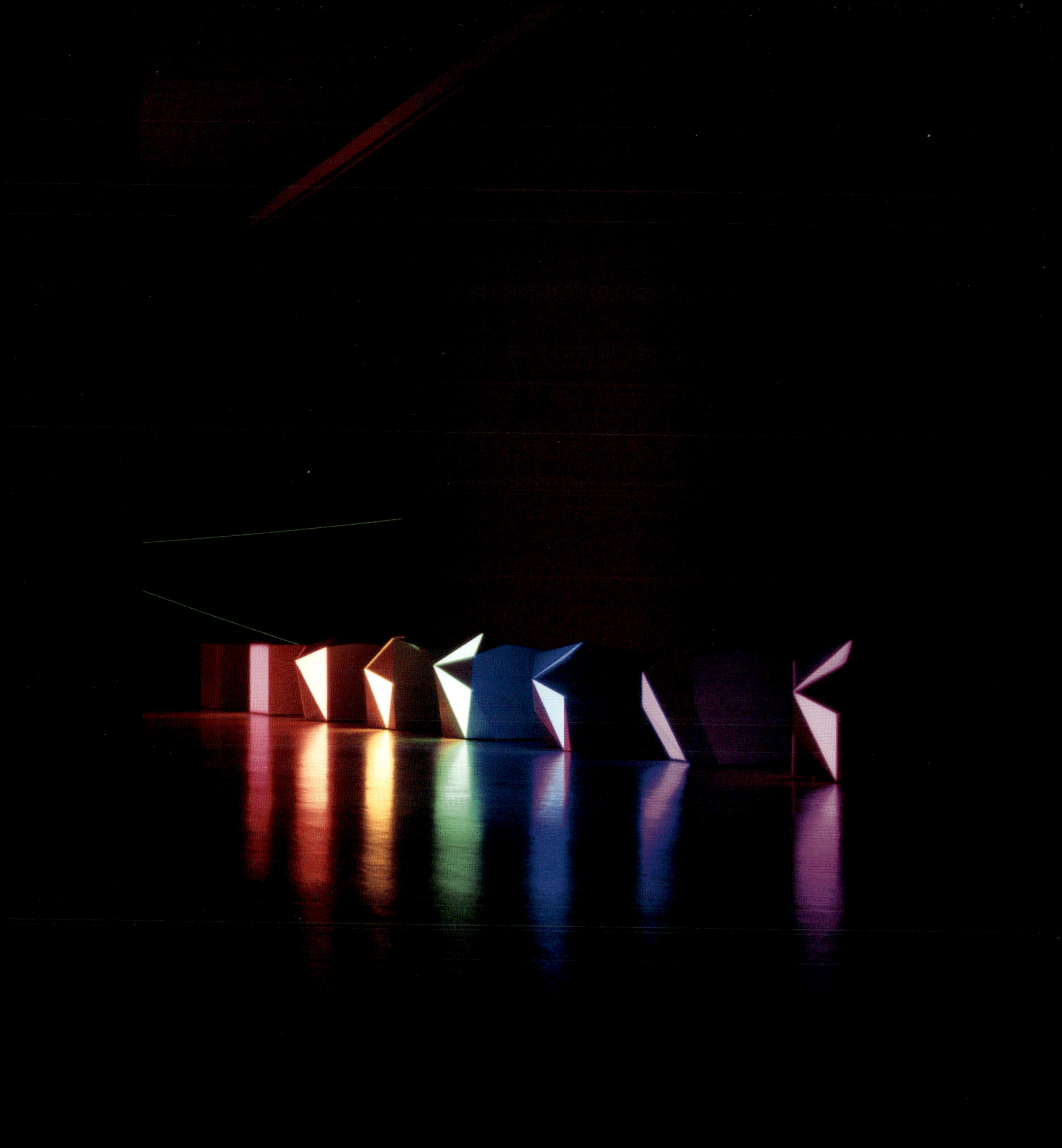

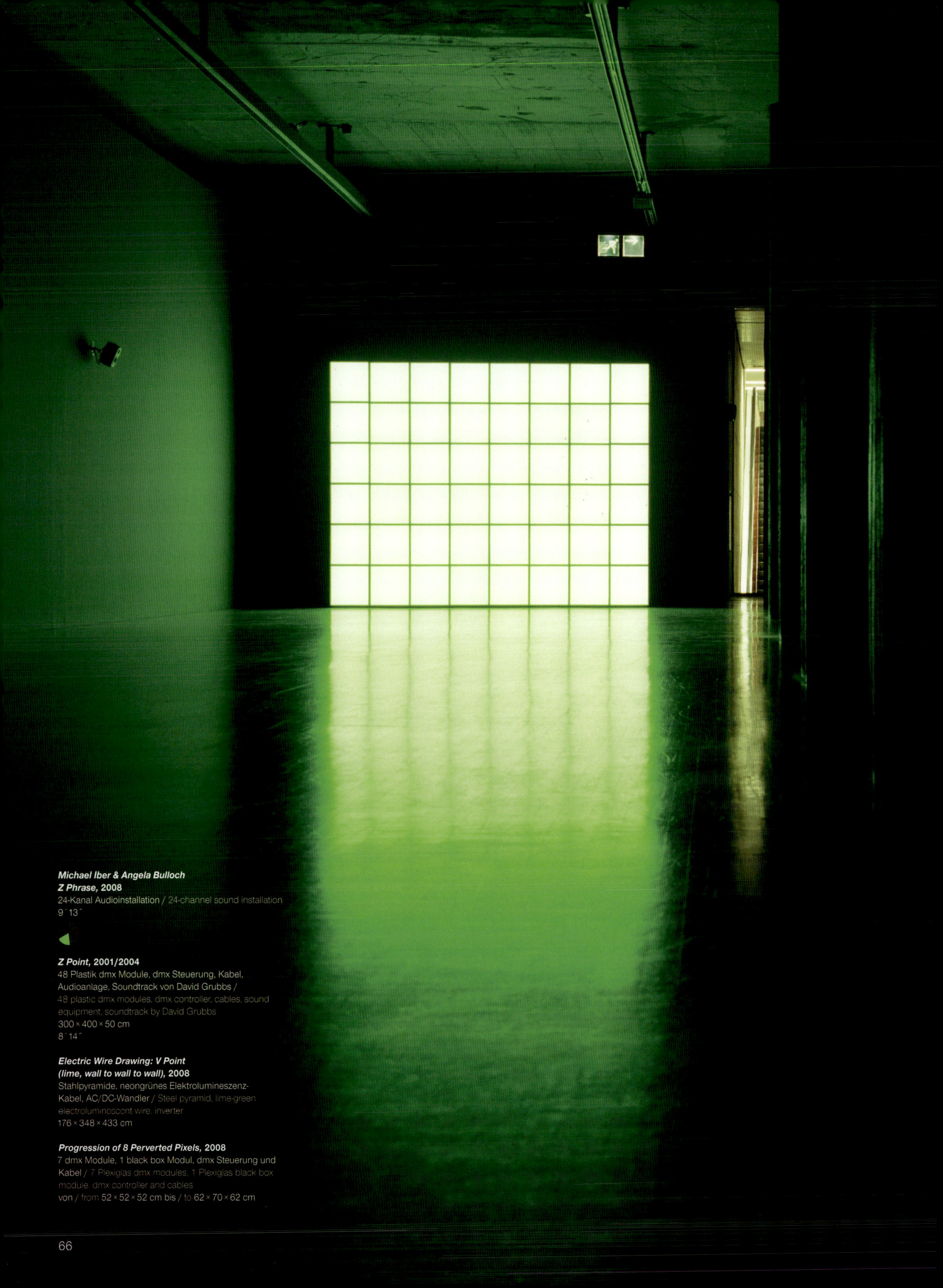

Michael Iber & Angela Bulloch
***Z Phrase*, 2008**
24-Kanal Audioinstallation / 24-channel sound installation
9´13˝

***Z Point*, 2001/2004**
48 Plastik dmx Module, dmx Steuerung, Kabel, Audioanlage, Soundtrack von David Grubbs / 48 plastic dmx modules, dmx controller, cables, sound equipment, soundtrack by David Grubbs
300 × 400 × 50 cm
8´14˝

***Electric Wire Drawing: V Point (lime, wall to wall to wall)*, 2008**
Stahlpyramide, neongrünes Elektrolumineszenz-Kabel, AC/DC-Wandler / Steel pyramid, lime-green electroluminescent wire, inverter
176 × 348 × 433 cm

***Progression of 8 Perverted Pixels*, 2008**
7 dmx Module, 1 black box Modul, dmx Steuerung und Kabel / 7 Plexiglas dmx modules, 1 Plexiglas black box module, dmx controller and cables
von / from 52 × 52 × 52 cm bis / to 62 × 70 × 62 cm

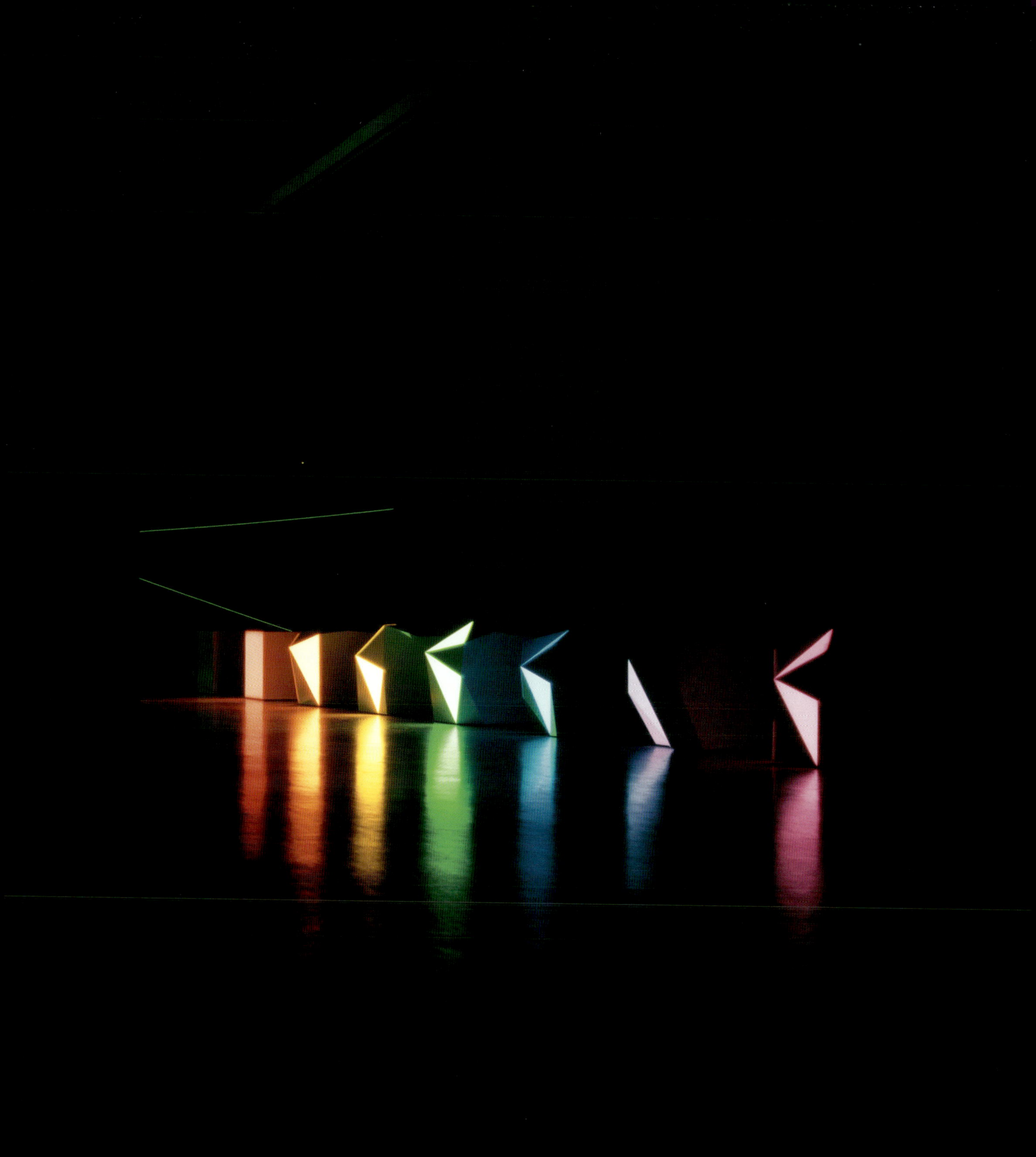

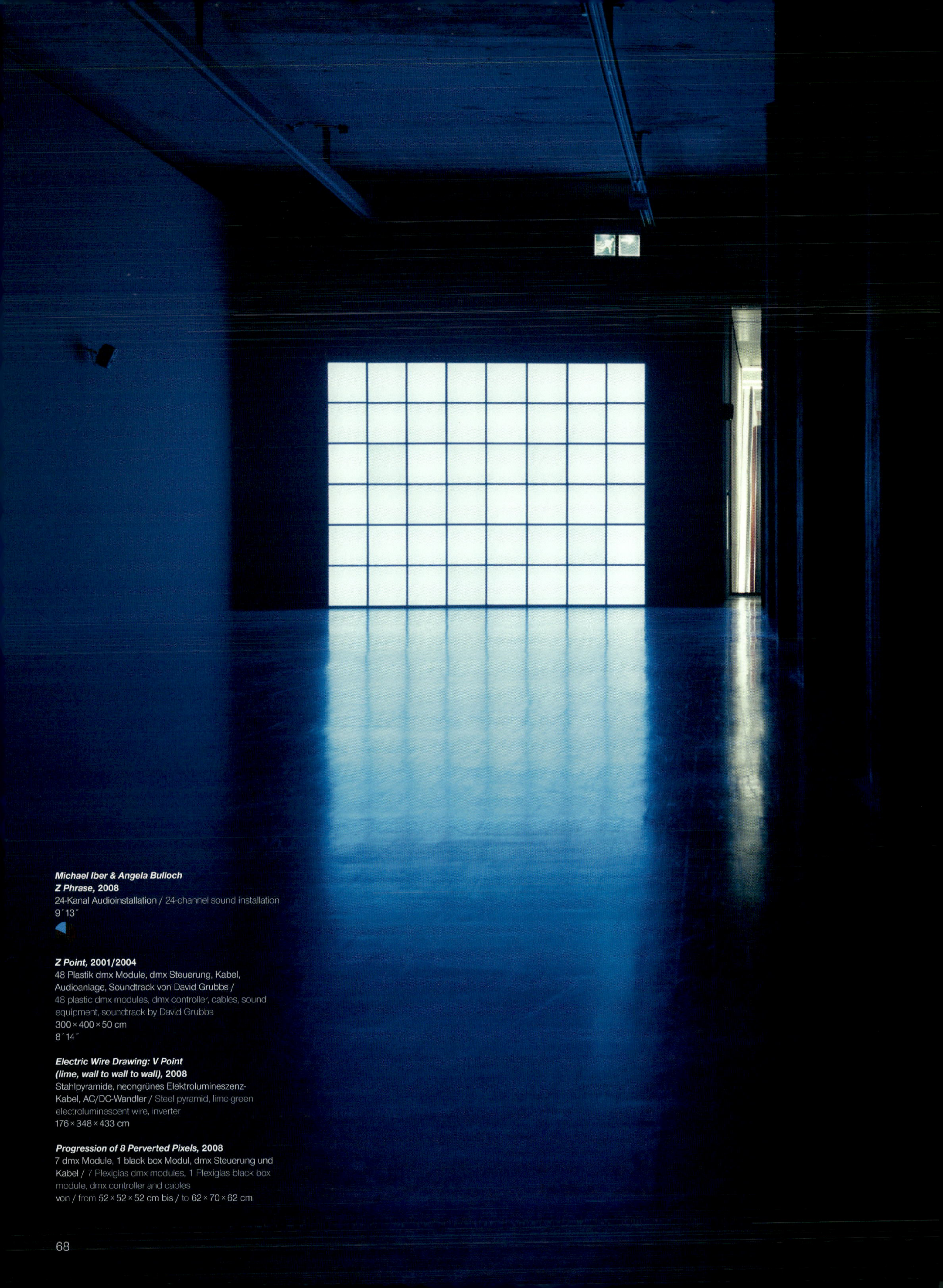

Michael Iber & Angela Bulloch
***Z Phrase*, 2008**
24-Kanal Audioinstallation / 24-channel sound installation
9´13˝

***Z Point*, 2001/2004**
48 Plastik dmx Module, dmx Steuerung, Kabel, Audioanlage, Soundtrack von David Grubbs / 48 plastic dmx modules, dmx controller, cables, sound equipment, soundtrack by David Grubbs
300 × 400 × 50 cm
8´14˝

***Electric Wire Drawing: V Point (lime, wall to wall to wall)*, 2008**
Stahlpyramide, neongrünes Elektrolumineszenz-Kabel, AC/DC-Wandler / Steel pyramid, lime-green electroluminescent wire, inverter
176 × 348 × 433 cm

***Progression of 8 Perverted Pixels*, 2008**
7 dmx Module, 1 black box Modul, dmx Steuerung und Kabel / 7 Plexiglas dmx modules, 1 Plexiglas black box module, dmx controller and cables
von / from 52 × 52 × 52 cm bis / to 62 × 70 × 62 cm

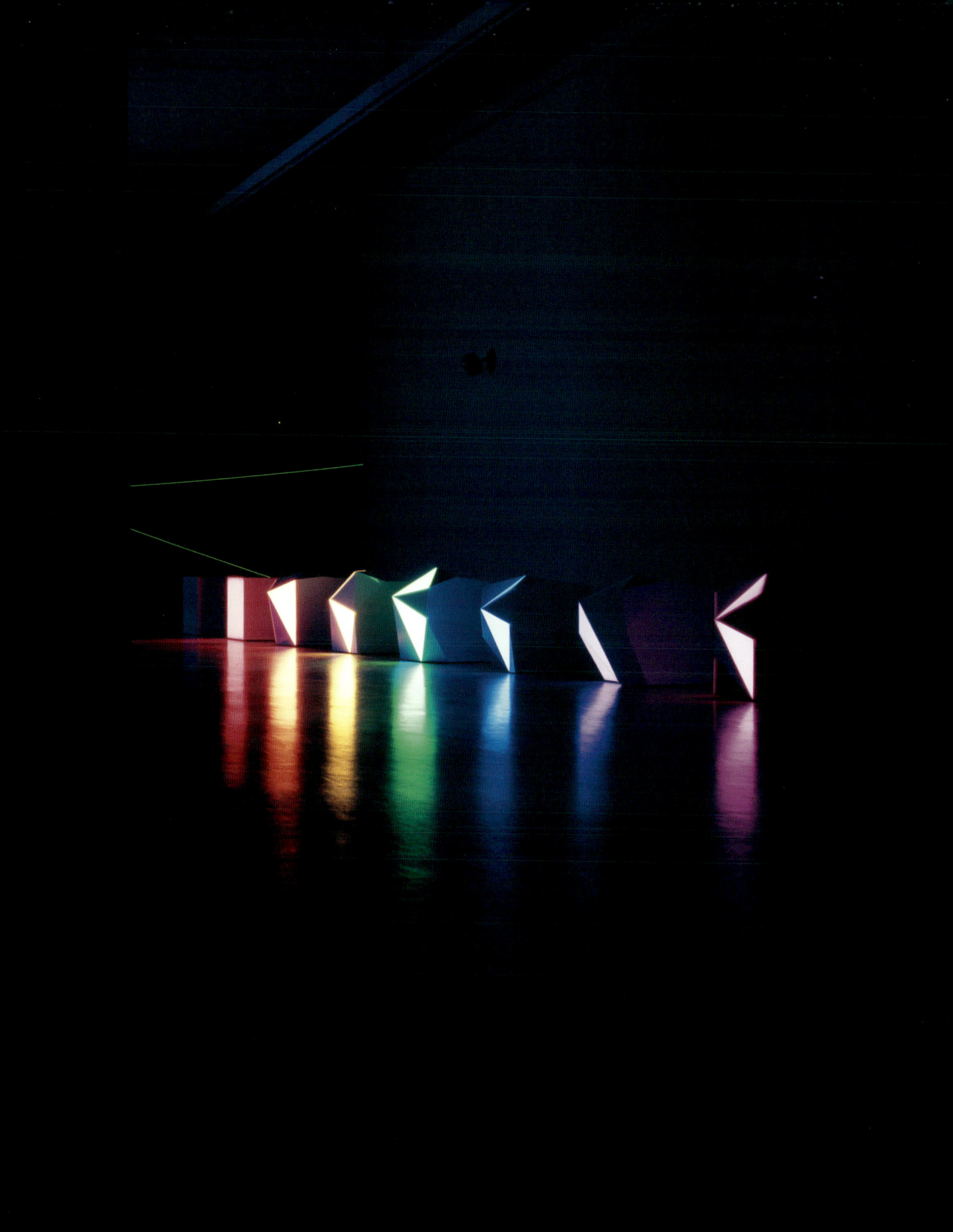

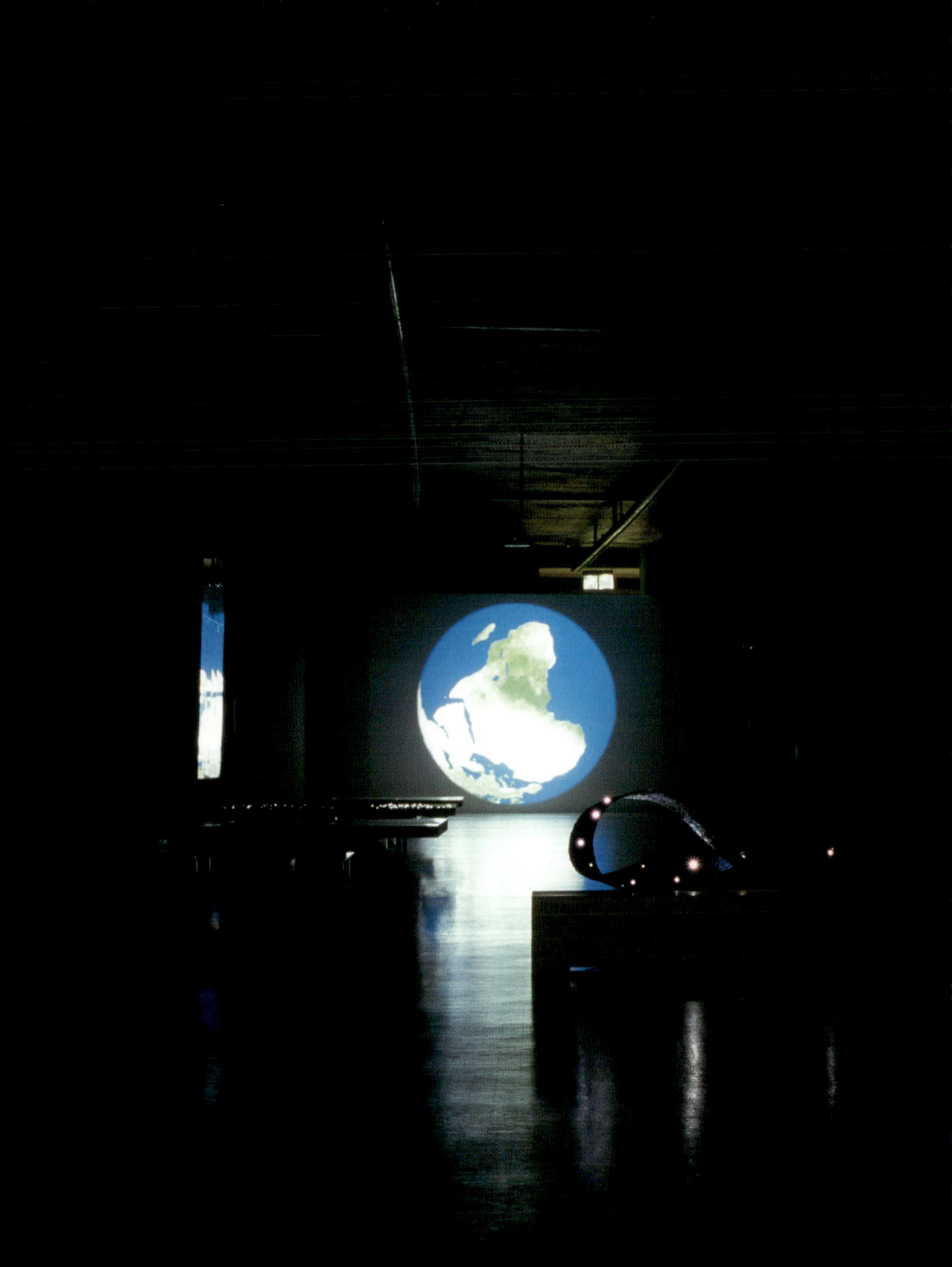

***Night Sky: E.T. from Mercury.12*, 2007**
LED-Installation, Neopren, Pollopasspiegel, Alu-Profile / LED installation, Neoprene, Pollopas mirror, aluminium profile
Boden / floor 266,4 × 199,8 × 45 cm,
Wand / wall 266,4 × 199,8 × 20 cm

***Night Sky: Mars from Venus.12*, 2007**
LED-Installation, Neopren, Pollopasspiegel, Alu-Profile / LED installation, Neoprene, Pollopas mirror, aluminium profile
Boden / floor 266,4 × 199,8 × 45 cm,
Wand / wall 266,4 × 199,8 × 20 cm

***Night Sky: Gleise 581C.12*, 2008**
LED-Installation, Neopren, Pollopasspiegel, Alu-Profile / LED installation, Neoprene, Pollopas mirror, aluminium profile
Boden / floor 266,4 × 199,8 × 45 cm,
Wand / wall 266,4 × 199,8 × 20 cm

***Last Year Again. 2008*, 2008**
Synchronisierte Doppelprojektion, digitale Videoanimation, 2 Mac minis, Videoprojektoren / Synchronised double projection, digital video animation, 2 Mac minis, video projectors
3´ 06˝

***Möbius Night Sky Model: Mark I*, 2003**
Stahl, Bitumen, LED-Installation, dmx Steuerung, Kabel und hölzerner Sockel / Steel, bitumen, LED installation, dmx controller, cables and wooden plinth
71 × 70 × 130 cm

***Electric Wire Drawing: F. S. Untitled (red, floor)*, 2008**
Stahl, rotes Elektrolumineszenz-Kabel, AC/DC-Wandler / Steel tubes, red electroluminescent wire, inverter
4,1 × 11,4 × 365,8 cm

***Electric Wire Drawing: F. S. Untitled (cyan, wall to floor)*, 2008**
Blaugrünes Elektrolumineszenz-Kabel, AC/DC-Wandler / Cyan electroluminescent wire, inverter
279 × 31 × 10 cm

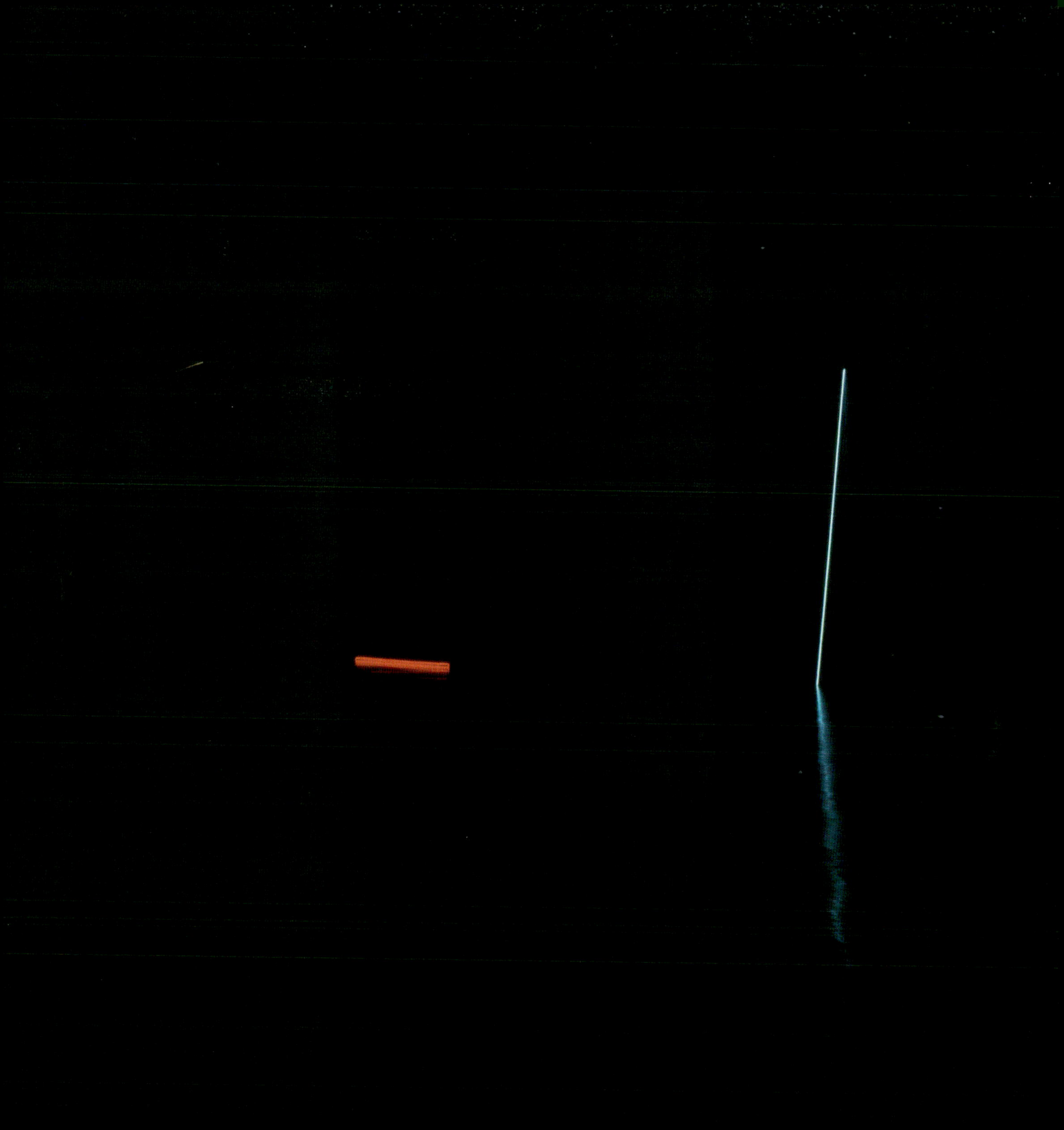

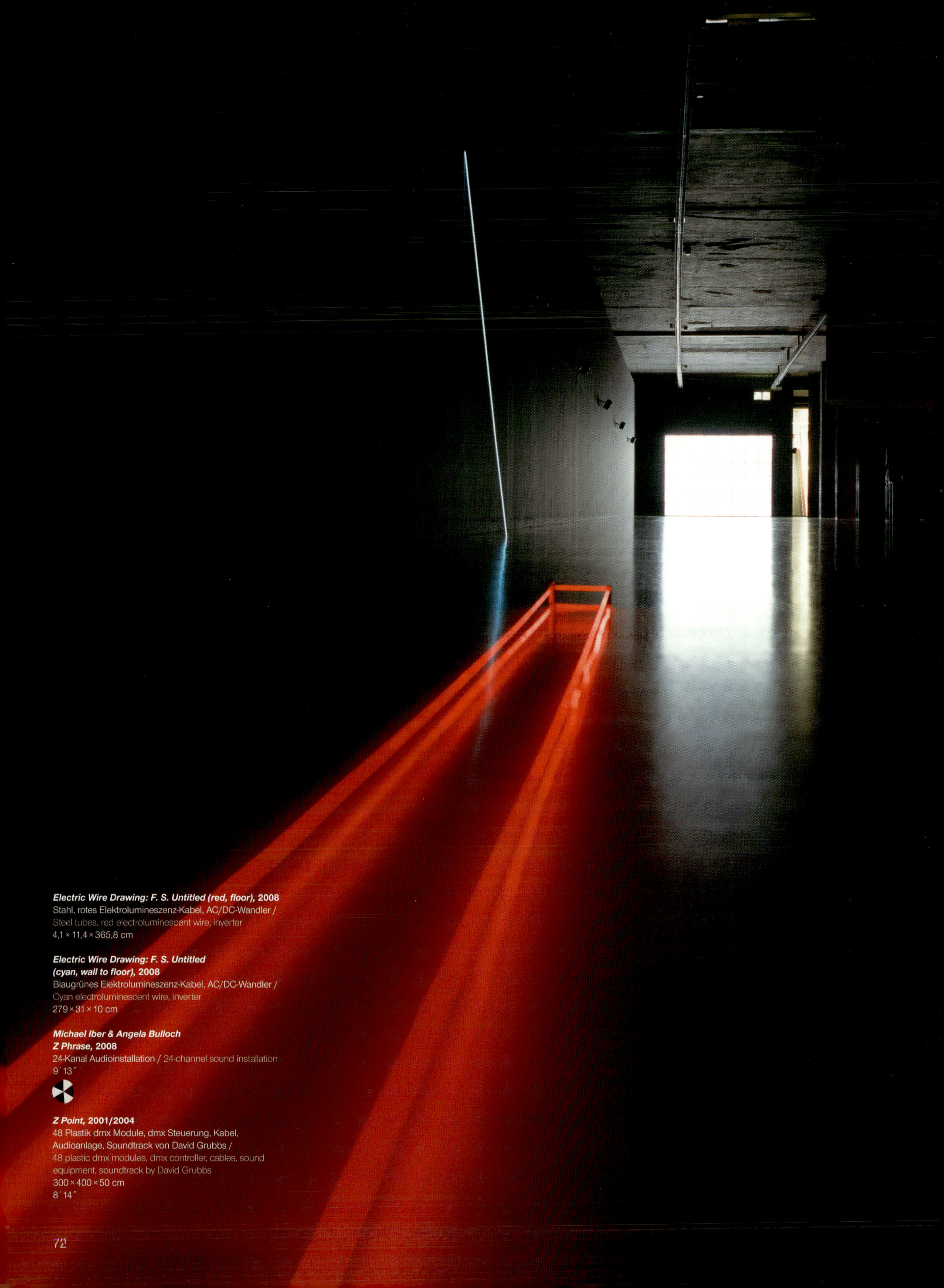

Electric Wire Drawing: F. S. Untitled (red, floor), 2008
Stahl, rotes Elektrolumineszenz-Kabel, AC/DC-Wandler /
Steel tubes, red electroluminescent wire, inverter
4,1 × 11,4 × 365,8 cm

Electric Wire Drawing: F. S. Untitled (cyan, wall to floor), 2008
Blaugrünes Elektrolumineszenz-Kabel, AC/DC-Wandler /
Cyan electroluminescent wire, inverter
279 × 31 × 10 cm

Michael Iber & Angela Bulloch
Z Phrase, 2008
24-Kanal Audioinstallation / 24-channel sound installation
9´13´´

Z Point, 2001/2004
48 Plastik dmx Module, dmx Steuerung, Kabel, Audioanlage, Soundtrack von David Grubbs /
48 plastic dmx modules, dmx controller, cables, sound equipment, soundtrack by David Grubbs
300 × 400 × 50 cm
8´14´´

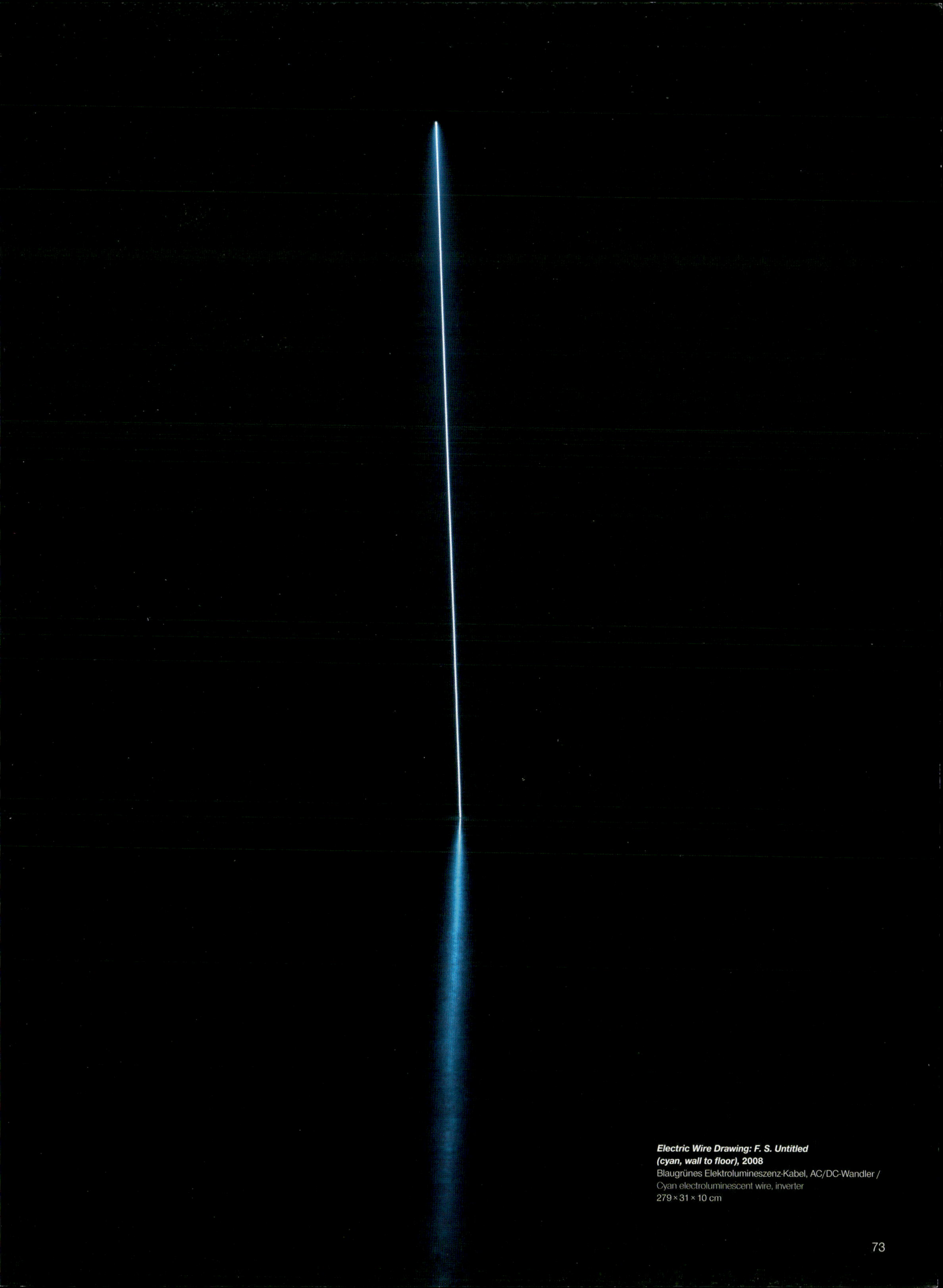

***Electric Wire Drawing: F. S. Untitled (cyan, wall to floor),* 2008**
Blaugrünes Elektrolumineszenz-Kabel, AC/DC-Wandler / Cyan electroluminescent wire, inverter
279 × 31 × 10 cm

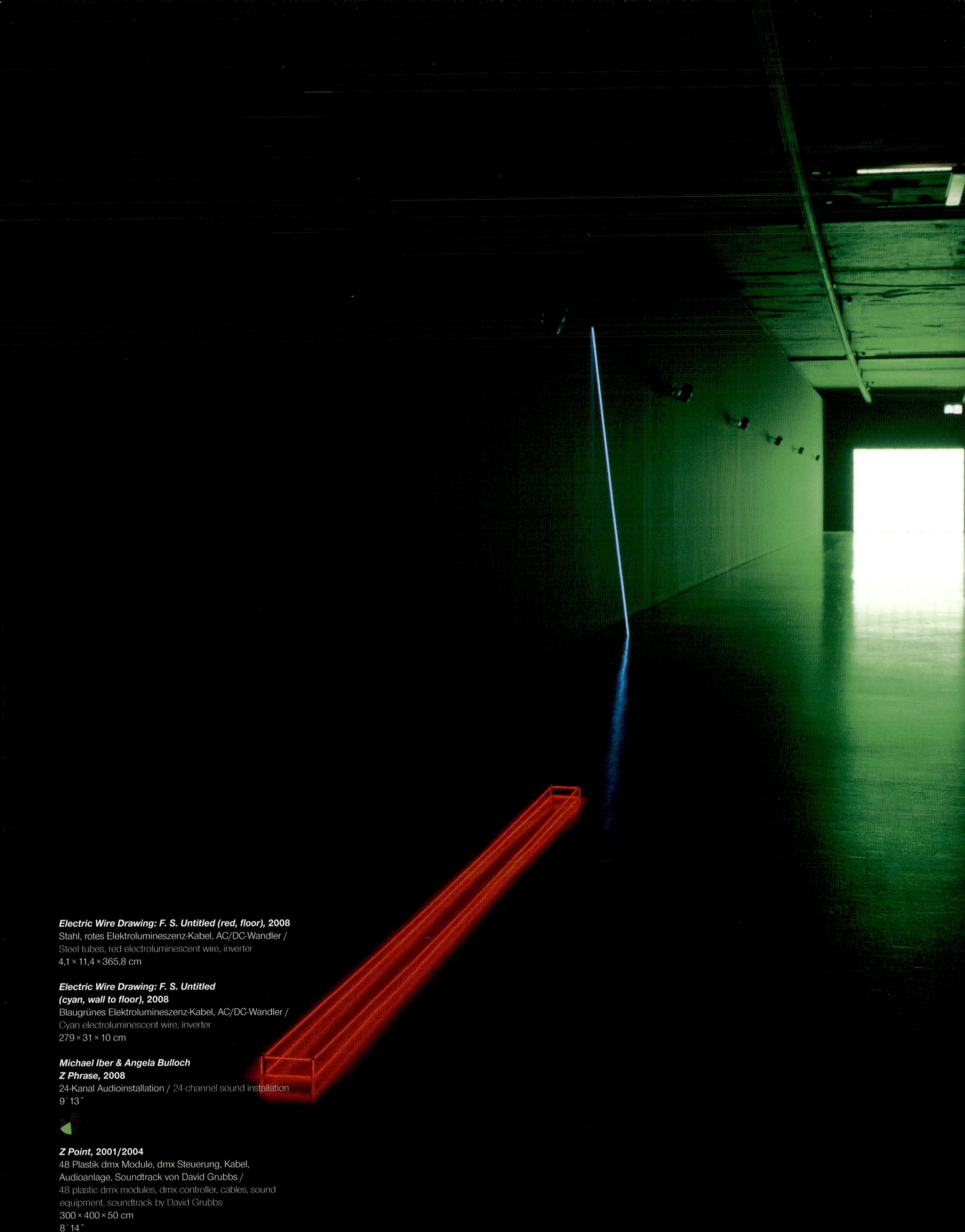

Electric Wire Drawing: F. S. Untitled (red, floor), 2008
Stahl, rotes Elektrolumineszenz-Kabel, AC/DC-Wandler /
Steel tubes, red electroluminescent wire, inverter
4,1 × 11,4 × 365,8 cm

Electric Wire Drawing: F. S. Untitled
(cyan, wall to floor), 2008
Blaugrünes Elektrolumineszenz-Kabel, AC/DC-Wandler /
Cyan electroluminescent wire, inverter
279 × 31 × 10 cm

Michael Iber & Angela Bulloch
Z Phrase, 2008
24-Kanal Audioinstallation / 24-channel sound installation
9´13˝

Z Point, 2001/2004
48 Plastik dmx Module, dmx Steuerung, Kabel,
Audioanlage, Soundtrack von David Grubbs /
48 plastic dmx modules, dmx controller, cables, sound
equipment, soundtrack by David Grubbs
300 × 400 × 50 cm
8´14˝

Z Phrase

Für die Ausstellung ***The SpaceThat Time Forgot*** bat Angela Bulloch den Musiker Michael Iber, zusammen mit ihr eine Soundinstallation zu konzipieren. Die 24 Kanal Audioinstallation ***Z Phrase*** ist auf einen einminütigen Ausschnitt aus dem von David Grubbs für Angela Bullochs ***Z Point*** entwickelten Soundtrack konzentriert. Mit Hilfe zweier von Michael Iber entwickelter Computerprogramme wird dieses Ausgangsmaterial vielfach variiert und im Raum verteilt: als Ganzes oder in zeitlich stark gedehnten kleinsten Partikeln, für sich alleinstehend oder zu Klangkaskaden aufeinander getürmt. Mit der im Rahmen einer langjährigen Zusammenarbeit mit dem Dirigenten Christian von Borries entstandenen ***Soundalike***-Software hat Michael Iber sich ein Werkzeug geschaffen, welches in der Lage ist, beliebige Audio-Dateien in Orchesterpartituren zu übersetzen. Die Besetzung des Orchesters lässt sich dabei ebenso frei definieren wie die zeitliche Rasterung, die ein wesentliches Kriterium für die Spielbarkeit der Partituren darstellt. Interessant ist an diesem Verfahren die Unschärfe zwischen dem Ausgangsmaterial und dem daraus gewonnenen neuen Stück: so tragen die ***Soundalikes*** eine starke Analogie zu den Reduktions- und Abstraktionstechniken von Angela Bulloch.
Entgegen ihrer ursprünglichen Konzeption, der Generierung spielbaren Orchestermaterials, setzten Angela Bulloch und Michael Iber ***Soundalike*** mittels eines neuen Software-Moduls nun erstmals als Grundlage klanglicher Synthese ein. Dabei hat der im Vergleich zu herkömmlichen Synthese-Resynthese-Verfahren eingeschlagene „Umweg" über die Partiturerstellung nicht nur klanglich sehr charakteristische Auswirkungen: er prädestiniert die Verteilung der Stimmen auf die einzelnen Lautsprecher im Raum.
Musik im Kopf heißt das zweite der erwähnten Computerprogramme und prägt den Mittelteil von ***Z Phrase.*** Der eigentümliche Name leitet sich von einer Äußerung des schon fortgeschritten schizophrenen Robert Schumann über seine Wahrnehmung musikalischer Zeit ab: dieses Software-Instrument ermöglicht extreme zeitliche Beschleunigungen und Dehnungen bis hin zum Stillstand sowie davon unabhängige Tonhöhenveränderungen im Echtzeit-Betrieb und erlaubt so ein investigatives Abtasten eines Klanges im Hinblick auf seine charakteristischen Momente.

Michael Iber wurde 1965 in Mannheim geboren und lebt in Berlin und Coburg. Er studierte Klavier an der Staatlichen Hochschule für Musik Karlsruhe, der California State University und der Royal Academy of Music in London und war bis 2005 als Pianist tätig. Seit Ende der 1990er Jahre beschäftigt er sich mit elektronischer Musik und programmiert die meisten der dafür benötigten Werkzeuge selbst.
Ursprünglich klassischer Interpret interessiert er sich inzwischen in einem erweiterten Sinne für Interpretationsformen, für ein verändertes Verständnis von Kunstwerk und Urheberschaft und die Bedeutung kultureller Referenzen im Rahmen künstlerischer Arbeit. Er ist Autor zahlreicher Rundfunk-Features und arbeitet für den SWR in der Produktion der Donaueschinger Musiktage.

Z Phrase

For the exhibition *The Space That Time Forgot,* Angela Bulloch asked musician Michael Iber to work with her on a sound installation. The resulting 24-channel piece *Z Phrase* focuses on a one-minute excerpt from David Grubbs' soundtrack for Angela Bulloch's *Z Point.* With the help of two computer programs developed by Iber, this source material is varied in many ways and distributed around the space: as a whole or as stretched fragments, standing on their own or piled up into cascades of sound.
The *Soundalike* software developed in cooperation with conductor Christian von Borries over a period of years is a tool capable of translating any audio file into an orchestra score. The instrumentation and the size of the orchestra can be freely defined, as can the resolution of the timescale, a key criteria for the playability of scores. An interesting aspect of this technique is the blurring that takes place between the source material and the resulting new piece, and in this sense, the *Soundalikes* bear a strong resemblance to the techniques of reduction and abstraction employed by Angela Bulloch.
Instead of using *Soundalike* for its original purpose of generating playable orchestra material, Angela Bulloch and Michael Iber added a new software module and used it as the basis for sound synthesis. The creation of scores—a detour in comparison with conventional synthesis/resynthesis techniques—has very characteristic results, not just in terms of sound: it also predestines the results for spreading across individual loudspeakers in the space.
The second computer programme, which shapes the middle section of *Z Phrase,* is called *Musik im Kopf* (Music In The Head). The name derives from a comment by Robert Schumann, made when he was already in an advanced state of schizophrenia, about his perception of musical time: this software tool allows extreme time-acceleration and -stretching to the point of stasis in real-time, as well as independent pitch-shiftings, making it possible to investigate a sound with regard to its characteristic moments.

Michael Iber (born in Mannheim in 1965) lives in Berlin and Coburg. He studied piano at Karlsruhe University of Music, California State University and London's Royal Academy of Music, working as a pianist until 2005. Since the late 1990s, he has been working on electronic music, programming most of the necessary tools himself.
Originally a classical player, he is now interested in forms of interpretation in a wider sense, for an altered understanding of the artwork and authorship and the meaning of cultural references in the context of artistic work. He has written numerous radio features and works for the South German broadcaster SWR on the production of the Donaueschingen Festival.

Matthias Mühling

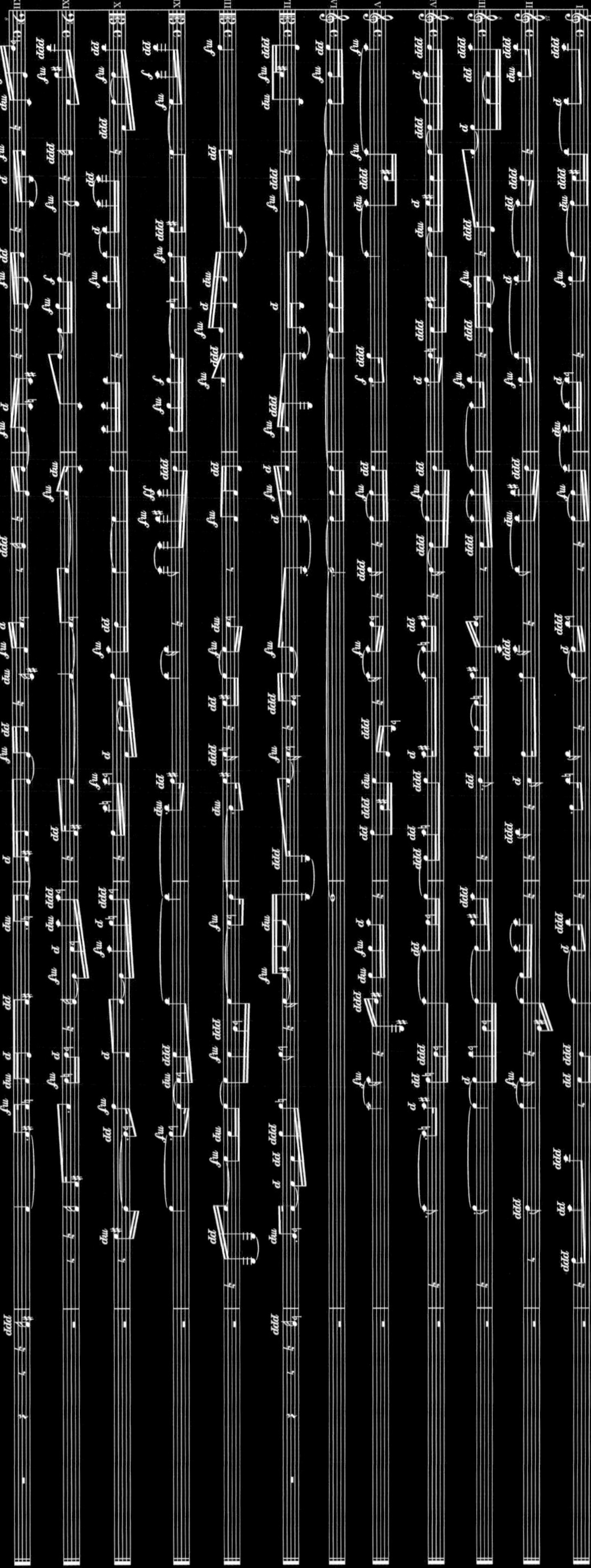

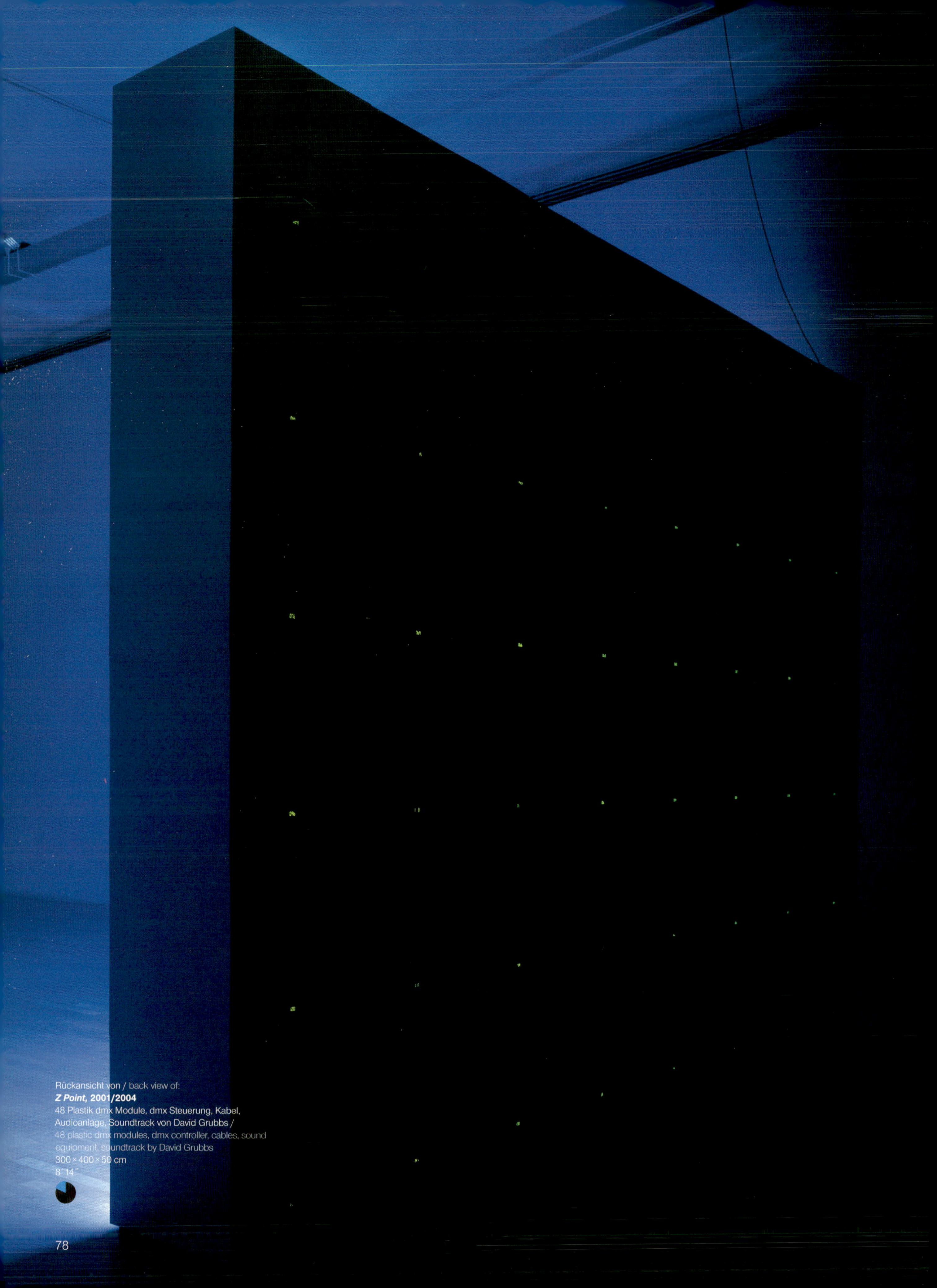

Rückansicht von / back view of:
***Z Point*, 2001/2004**
48 Plastik dmx Module, dmx Steuerung, Kabel, Audioanlage, Soundtrack von David Grubbs /
48 plastic dmx modules, dmx controller, cables, sound equipment, soundtrack by David Grubbs
300 × 400 × 50 cm
8′ 14″

Projektion und Perspektive

Angela Bullochs Ausstellung im Kunstbau des Lenbachhauses

Matthias Mühling

Last Year Again

Das für die Bewohner unseres Planeten wohl mit dem größten Identifikationspotential ausgestattete Bild der Erde ist ein Anblick, den fast niemand jemals real gesehen hat. Einigen wenigen Astro- und Kosmonauten ist dieser Blick bisher vergönnt gewesen, trotzdem hat diese Ansicht nachhaltig die Vorstellung aller Menschen von der Erde beeinflusst. Angela Bulloch präsentiert dieses Bild als digitale Animation der Erde in Bewegung. Sie rotiert in unterschiedlichen Geschwindigkeiten und zeigt das unwahrscheinliche Phänomen eines global wolkenlosen Himmels. Als Doppelprojektion angelegt sind die Bilder von *Last Year Again* seltsam vertraut und doch irreal als reales Erlebnis. Es ist die Perspektive, die eben keine menschliche ist und die als Perspektive aus dem Weltraum zurück auf die Erde das Prinzip der Erde als Körper überhaupt erst fassbar macht. Denn auf der Erdoberfläche erleben wir unseren Planeten eben doch als eine Scheibe, auf der wir umherlaufen. Und so, wie der Fortschritt in Wissenschaft und Technik unseren Blick auf die Umwelt erweitert hat, so zeigt Angela Bulloch mit eigentlich allen Arbeiten in der Ausstellung, wie wenig davon Erleben und wie viel daran Projektion sein kann.

Z Point

Paradigmatisch ist dies in der Arbeit *Z Point* von 2001/2004 verwirklicht. Es ist eine Arbeit aus 48 sogenannten Pixel Boxen, die von Angela Bulloch entworfen wurden. Die Pixel Box vergrößert die kleinste, eigentlich unsichtbare Einheit des digitalen Bildes auf die Größe einer minimalistischen Skulptur. Das Innere der Pixel Box besteht aus einem RGB-Lichtsystem und kann 16 Millionen Farben generieren. Mittels eines DMX-Modulsystems können die Boxen untereinander verbunden und mit jeder möglichen Bildinformation programmiert werden. *Z Point* bezieht sich auf die letzten 8 Minuten des von Michelangelo Antonioni 1970 realisierten Films *Zabriski Point,* also auf jenen Moment, in dem Daria Halprin die Explosion einer modernistischen Luxusvilla beobachtet oder aber imaginiert. Die kinematographische Disposition des Films aus 24 Bildern pro Sekunde wird in *Z Point* auf einen Lichtwechsel pro Sekunde ausgedehnt. Es ist ein bewusstes Zurücktreten hinter die technischen Möglichkeiten digitaler Bilder: Entschleunigung der Sequenzen auf wahrnehmbare Lichtwechsel und radikale Verringerung der Anzahl der Pixel. Dies führt zu einer bewussten Darstellung des quasi Apparatischen der verwendeten Technik. Nur der Soundtrack verweist noch auf den ungebrochenen narrativen Fluss der filmischen Vorlage. In der Ausstellung *The Space That Time Forgot* wird *Z Point* mit einem von David Grubbs komponierten Soundtrack gezeigt. *Z Point* mit einer Laufzeit von 8 Minuten und 14 Sekunden befindet sich

Projection and Perspective

Angela Bulloch's Exhibition at Lenbachhaus Kunstbau

Matthias Mühling

Last Year Again

The image of the Earth with the greatest potential for identification for our planet's inhabitants is the view which almost nobody has ever really seen. A few astronauts and cosmonauts have been granted this sight, but despite that, this view has indelibly affected the idea that everybody has about the Earth. Angela Bulloch presents this image as a digital animation of the Earth in motion. It rotates around its own axis and shows the improbable phenomenon of a globally cloudless sky. Designed as a double projection, the images of *Last Year Again* are strangely familiar, but nevertheless unreal as a real experience. It is an extra-terrestrial perspective that makes the principle of the Earth as a solid body graspable. Because we are on the Earth's surface, we experience our planet simply as a flat disk upon which we walk. And so, just as progress in science and technology has expanded our view of the surrounding world, so Bulloch shows in almost all the works exhibited just how much of this view is experienced and how much can be projection.

Z Point

This idea is realized paradigmatically in *Z Point of* 2001–04. It is a work composed of forty-eight pixel boxes. The pixel box enlarges the smallest, and usually invisible, unit of the digital image to the size of a Minimalist sculpture. The insides of the box consist of an RGB light system and can generate sixteen million colours. By means of a DMX module system, the boxes can be interconnected and programmed with any kind of image information. *Z Point* refers to the last eight minutes of Michelangelo Antonioni's 1970 film, *Zabriski Point,* when Daria Halprin observes or imagines the explosion of a modernist, luxury villa. In *Z Point,* the cinematographic film's sequence of twenty-four images per second is reduced to one change per second. It is a conscious stepping-back behind the technological potential of digital images, a deceleration of the sequences to perceptible changes of light and a radical reduction of the number of pixels. This leads to an intentional reference to the technology employed. Only the sound track still refers to the unbroken narrative flow of the filmic script. In the exhibition, *Z Point* is displayed with a sound track composed by David Grubbs. With a duration of eight minutes and fourteen seconds, *Z Point* is normally endlessly looping, but for the exhibition at Lenbachhaus Kunstbau, it is synchronized with the twenty-four-channel sound installation *Z Phrase.* This sound installation has been especially developed by Bulloch and the musician Michael Iber for the subterranean space of the Kunstbau and has a duration of nine minutes and thirteen seconds. Both *Z Point* and *Z Phrase* are played alternately three times per hour. Whenever we hear *Z Phrase* running, the sound of *Z Point* is switched off and the monitors show a uniformly red, green, or blue surface.

normalerweise im endlosen Loop. In der Ausstellung im Kunstbau des Lenbachhauses ist die Installation jedoch mit der 24 Kanal Audioinstallation *Z Phrase* synchronisiert. Diese Audioinstallation wurde von Angela Bulloch und dem Musiker Michael Iber speziell für den unterirdischen Raum des Kunstbaus entwickelt und hat eine Länge von 9 Minuten und 13 Sekunden. *Z Point* wie *Z Phrase* werden abwechselnd drei Mal pro Stunde gespielt. Während wir *Z Phrase* hören können, ist *Z Point* abgeschaltet und seine 48 Pixel Boxen zeigen entweder eine einheitlich rote, grüne oder blaue Oberfläche, entsprechend den drei Grundfarben der Leuchtstoffröhren, die sich im Innern der Boxen befinden.

Was jedoch die Qualität der Arbeit ausmacht, ist die Spannung zwischen diesen Komponenten: der technischen Reflexion digitaler Bildwelten und den semantischen Konnotationen. Diese werden einerseits durch die filmische Vorlage und andererseits durch die formale Gestaltung der Pixel Boxen in Anlehnung an die Minimal Art ausgelöst. So trifft sich auf der semantischen Ebene das Thema des Films mit der Abstraktionsleistung der Pixel Box. Während im Film nicht zu entscheiden ist, ob Daria Halprin die Explosion der Villa nur imaginiert oder ob sie tatsächlich explodiert, kann der Bilderfluss der Pixel Boxen als wechselnde abstrakte Farbfelder oder als grob aufgelöste Bilder des Filmes gelesen werden. Der Betrachter befindet sich mit seiner Wahrnehmung zwischen den möglichen Varianten. Er entscheidet, ob *Z Point* verbindliche Beziehungen zwischen Abbild und Vorbild, Zeichen und Bezeichnetem entweder dementiert oder anbietet.

Progression of 8 Perverted Pixels

Während die Pixel Box im Kontext von *Z Point* noch als modulares Gestaltungselement eingesetzt ist, wird dieses gleichberechtigte Nebeneinander der Gestaltungselemente in *Progression of 8 Perverted Pixels* von 2008 aufgegeben. Mit der kristallinen Aufsplitterung ihrer eigenen Syntax, der Pixel Box, überschreitet Angela Bulloch den vermeintlichen Endpunkt progressionsorientierter Kunstgeschichtsschreibung. So wird der ausgetretene Pfad des Reduktions- und Abstraktionsinteresses der Moderne verlassen, welcher einen seiner vielen Kulminationspunkte in der Reduktion auf die Form des Würfels in der Minimal Art gefunden hatte. Die Reihe aus acht sukzessive gesteigerten Pervertierungen, Fragmentierungen oder prismatischen Aufspaltungen transformiert die solide und einfache Form des Würfels in immer komplexere geometrische Formen, die mathematischen Regeln folgen. Orientiert an der Regelhaftigkeit der platonischen Formen zersplittert die Pixel Box in ein vielgestaltiges Objekt, dessen perspektivische Aufspaltung dem geregelten und einfachen Volumen des Würfels entgegensteht.

The interest of *Z Point* consists in the tension between technical reflection and the semantic connotations of the analogue and the digital, of the filmic and the sculptural—of medium and form. The latter are triggered by the filmic script and by the grid of the pixel boxes that evoke Minimal Art. Thus, on a semantic level, the film's subject encounters the abstraction achieved by the pixel box. Whereas in the film it cannot be decided whether Daria Halprin only imagines the explosion of the villa or whether it really does explode, the flow of images in the pixel boxes can be read either as changing, abstract, coloured squares or as roughly dissolved images of the film. In their perception, viewers are situated between two possible variants. They decide whether *Z Point* either denies or offers reliable relations between portrayal and portrayed, between sign and signified.

Progression of 8 Perverted Pixels

Whereas in the context of *Z Point* the pixel box is still used as a modular element, this democratic coexistence of each element is given up in *Progression of 8 Perverted Pixels* of 2008. With the pixel box's crystalline splitting of its own syntax, Bulloch transgresses the supposed end point of an art history premised upon the notion of artistic progression. Thus, the artist leaves the beaten track of modern art's interest in reduction and abstraction, which had found one of its many points of culmination in Minimal Art's reduction of form to the cube. The row of eight, successively intensified perversions, fragmentations, or prismatic splittings transforms the simple cube into increasingly complex forms that follow mathematical rules. Oriented toward the regularity of Platonic solids, the pixel box splits the image into a polymorphic object whose triangular facets and broken perspective contradict the simple volume of the cube.

Fly Over Genoa and *Flyover Genoa*

The problem of perspective in relation to space and its perception is taken up in the exhibition in a modified form with an early work from 1992. *Fly Over Genoa* and *Flyover Genoa* take advantage of a peculiarity in the English language, that is, that the two words "fly" and "over" change their meaning when the blank space between them is deleted to form a composite noun. "Flyover" means a section of an elevated road that avoids crossings and traffic lights and cuts a harsh path through the urban landscape. As the dream of the automobile city dreamt by urban planners in the 1960s and 1970s, these elevated roads have developed into an anti-urban nightmare. Initially, *Fly Over Genoa* and *Flyover Genoa* engage with this change of meaning in its immediate, literal sense. Whereas *Fly Over Genoa* has a flight over Genoa as its subject, *Flyover Genoa* shows a car ride on a flyover in Genoa. The first video, *Fly Over Genoa,* presents the view from the window of a

Fly Over Genoa und _Flyover Genoa_

Das Problem der Perspektive in Bezug auf den Raum und dessen Wahrnehmung wird in der Ausstellung in abgewandelter Form in einer Videoarbeit von 1992 zum Thema, welche für den Medienraum des Kunstbaus 2008 überarbeitet wurde. *Fly Over Genoa* und *Flyover Genoa* machen sich die Eigenschaft zu Nutze, dass im Englischen die beiden Wörter Fly und Over die Bedeutung wechseln, wenn ihnen als Kompositum das Leerzeichen zwischen ihnen entzogen wird. Flyover bezeichnet dann eine Hochstraße, die nur für den Autoverkehr zugänglich ist und Kreuzungen und Ampeln vermeidet, in dem zumeist auf Stelzen eine brutale Schneise durch die urbanen Stadtlandschaften geschlagen wird. Als Traum autogerechter Städte der Stadtplanung der 1960er und 1970er Jahre haben sich diese Hochstraßen zum anti-urbanen Alptraum entwickelt. Mit Leerzeichen bedeutet Fly over einfach, dass man über etwas hinwegfliegt. Die Arbeiten *Fly Over Genoa* und *Flyover Genoa* beschäftigen sich mit diesem Bedeutungswechsel zunächst in ihrem direkten Literalsinn. Während *Fly Over Genoa* ein Überflug von Genua zum Thema hat, zeigt *Flyover Genoa* eine Autofahrt auf der Hochstraße in Genua. Der erste Film (*Fly Over Genoa*) zeigt den Kamerablick aus dem Fenster einer Maschine der British Airways, der eben genau jenen Flyover in Genua zeigt, der im zweiten Film (*Flyover Genoa*) vom Rücksitz eines Auto heraus gefilmt wird. Der Flyover, auf welchem einerseits entlanggefahren und der andererseits überflogen wird, ist die Strada Aldo Moro in Genua, welche von Westen und Osten her entlang der Küste direkt in die Innenstadt von Genua führt. Da der Flug des ersten Films ein Landeanflug auf den Flughafen von Genua ist, kann die Autofahrt als Fortsetzung des Flugs gelesen werden, da die Strada Aldo Moro auch diejenige Strecke markiert, die ein Reisender vom Flughafen in die Stadt nehmen würde. Zudem wird diese quasi narrative Komponente der beiden Filme noch erweitert, wenn die Autofahrt über den Flyover in Richtung Innenstadt vor der Galleria Locus Solus endet, in der Angela Bulloch 1992 eine Einzelausstellung ausgerichtet wurde und in der *Fly Over Genoa* und *Flyover Genoa* zum ersten Mal gezeigt wurden. Die Kombination der beiden Filme kann also als eine Geschichte des Weges gelesen werden, den die Künstlerin genommen hat, um dorthin zu gelangen, wo das, was sie auf dem Weg dorthin produziert hat, in den Status eines Kunstobjekts erhoben wird.

Die Referenz auf den Christdemokraten Aldo Moro, in dessen Gedenken die Straße ihren Namen erhielt, eröffnet einen zusätzlichen Themenkomplex der Arbeit. Aldo Moro war der vor Berlusconi am längsten amtierende Regierungschef in Italien nach dem zweiten Weltkrieg (1963–1968 und 1974–1976). Am 16. März 1978 wurde er von den Roten Brigaden (*Brigate Rosse*) entführt und am 9. Mai 1978, nach 55-tägiger Geiselhaft, tot im Kofferraum eines Autos aufgefunden. Mario Moretti, der Anführer der Roten

British Airways flight showing the very same flyover in Genoa that is filmed in the second video, *Flyover Genoa,* from the back seat of a car. The flyover that is both driven along and flown over is the Strada Aldo Moro, which runs east-west along the coast and directly into the center of Genoa. Since the flight recorded in the first video is a landing at Genoa Airport, the car ride can be regarded as the continuation of the flight, because the Strada Aldo Moro also designates the route a traveler would take from the airport to the city. This narrative component of the two videos is further elaborated when the car ride on the fly-over in the direction of the inner city ends before the Galleria Locus Solus, where *Fly Over Genoa* and *Flyover Genoa* were exhibited in 1992. The combination of the two videos can therefore be read as a documentation of the artist's traveling to the institution, where, very significantly, her gesture of displaying a documentation of this trip takes on the status of an art object.

The reference to the Christian Democrat, Aldo Moro, in whose memory the road was named, adds an additional layer of complexity to the work. Aldo Moro was head of the government in post-war Italy. On March 16, 1978, he was kidnapped by the Brigate Rosse (Red Brigades) and found dead in the trunk of a car on May 9, 1978, after fifty-five days of captivity. Mario Moretti, the leader of the Red Brigades, had murdered him with eight shots after demands to release imprisoned members of the Red Brigades had not been fulfilled. With the death of Aldo Moro as a unifying figure, the historic process of reconciliation between Communists and Christian Democrats also came to an end. Several films, for example, *Il caso Aldo Moro (I giorni dell'ira)* directed by Giuseppe Ferrara in 1986, have also taken on this subject, and various theories have been formed about the background to the murder, all of which refer to sinister connections among international secret services, the Catholic Church, and the Mafia.

At the exhibition in Genoa, the videos were displayed together with a map on which both the flight path and the car ride were inscribed—a line, the primal form of drawing and, simultaneously, the marking of a path on the instrument associated with the invention of painting: the map. The cartographic style of the maps that Bulloch selected is also very painterly, because it tries to graphically represent the topographical formations, the fissured slopes in and around Genoa, as foothills of the mountain range. The work describes how one's perception of spaces and places depends upon the vehicle in which one finds oneself. Genoa is described as a city that draws its pride from its role as a former seafaring nation that opened up and dominated large expanses of the Earth, but also as a city that has lost its access to the sea through the separating power of roads, because the Strada Aldo Moro prevents unhindered access to the sea. The flyover is a dividing stream that accepts the price of the alienation of spaces for the benefit of mobility

Brigaden, hatte ihn mit acht Schüssen ermordet, nachdem den Forderungen nach Freilassung inhaftierter Mitglieder der Roten Brigaden nicht nachgekommen worden war. Mit dem Tod der Integrationsfigur Aldo Moro kam auch die historische Aussöhnung zwischen Kommunisten und Christdemokraten an ihr Ende. Mehrere Verfilmungen haben sich dieses Themas angenommen, und um die Hintergründe der Ermordung entspinnen sich unterschiedliche Theorien, welche sich alle auf die unheilvolle Verbindung zwischen internationalen Geheimdiensten, katholischer Kirche und der Mafia beziehen.
In der Ausstellung in Genua waren die Filme zusätzlich durch eine Landkarte flankiert, auf welcher sowohl die Flugbahn als auch die Autofahrt eingezeichnet sind. Eine Linie als Urform der Zeichnung und gleichzeitig als Markierung des Weges auf demjenigen Instrument, mit dem die Erfindung des Gemäldes in der europäischen Kunstgeschichte nördlich der Alpen assoziiert ist: der Landkarte. Der kartographische Stil der Landkarte, die Angela Bulloch ausgewählt hat, ist zudem ein sehr malerischer, da er die Geländeformationen, die zerklüfteten Hänge in Genua und um Genua herum, geradezu plastisch als Ausläufer des Gebirges vorzustellen sucht. Somit beschreibt die Arbeit auf ganz unterschiedliche Weise die Wahrnehmung von Räumen und Orten, je nach dem in welchem Verkehrsmittel man sich befindet. Gleichzeitig wird Genua als eine Stadt beschrieben, die ihren Stolz auf die geopolitische Eröffnung und Beherrschung der Großräume der Erde aus der Kontrolle und Praktizierung der Seefahrt bezieht, aber den Zugang zum Meer durch die trennende Kraft einer Straße eingebüßt hat. Denn die Strada Aldo Moro unterbindet den ungehinderten Zugang der Menschen zum Meer. Ein teilender Strom, welcher zu Gunsten der Mobilität und deren ökonomischer Bewegung die Entfremdung der Räume in Kauf nimmt. Somit sind in der Arbeit die unheilvollen Verbindungen von Herrschaft und Raum, Ökonomie und Mobilität auf vielfältigste Weise angesprochen.

Night Skies

Mithilfe eines für jedermann zugänglichen Computerprogramms hat Bulloch Ansichten bestimmter Ausschnitte des Weltalls erschaffen, die nicht mehr am Standpunkt des Erdbewohners orientiert sind. Es sind Blickrichtungen von bekannteren Planeten, wie Merkur, oder unbekannteren, wie Gleise 581 C im Sternbild Waage. Der wohl intelligenteste Anspruch der Arbeiten von Angela Bulloch in Bezug auf die Simulation dieser extraterrestrischen Perspektiven auf den Nachthimmel ist, dass sie jenseits dessen, was Repräsentation in der Kunst bedeuten konnte, ein gänzlich neues Phänomen erschließt. Während die Kunst immer Dinge generieren konnte, die es schlechthin nicht gibt oder gegeben hat, zeigt Angela Bulloch uns Dinge, die uns einerseits vertraut erscheinen, weil

and economic movement. In the work, therefore, the inauspicious connections between authority and territory, and between economy and mobility, are addressed in diverse ways.

Night Skies

With the aid of a common computer program, Bulloch has chosen snapshots of certain views of outer space that are no longer taken from the viewpoint of the Earth dweller. Probably the most intelligent claim of Bulloch's works in relation to the simulation of these extra-terrestrial perspectives on the night sky is that the simulation opens up a completely new phenomenon beyond what representation in art could mean. Whereas art has always been able to generate things that do not exist or have never existed, Angela Bulloch shows us things which on the one hand seem familiar, because they resemble what we can see in the sky on a clear night, but which on the other hand are clearly different from what we will ever be able to see, given the restrictions to our mobility and time. This excludes these views as representations of a presence. The *Night Skies* are based on algorithms and commands. To put it radically, Bulloch not only takes leave of a specific thinking of representation, but, at the same time, she also inaugurates an aesthetics concept that is based on algorithms generated by software of a antecedent hardware.

Electric Wire Drawings

Analogous to the *Night Skies,* two-dimensional images of multi-dimensional spaces in infinite outer space, Bulloch's *Electric Wire Drawings* designate real bodies, but rather than being present as solid masses in the exhibition space, they are represented only by means of their contours. The *Electric Wire Drawings* make direct reference to Fred Sandback by appropriating two works from his oeuvre. Bulloch transforms what was simply yarn or string in Sandback's works into luminous wires, which are lamp filaments in a linear form. Simple, straightforward objects, which were originally charged with a minimum of meaning, here become objects that acquire meaning through reference to Minimal Art while at the same time having always been over-determined by history through the choice of their form as Platonic solids.

Fred Sandback (1943–2003) is among the central figures of Minimal Art. His frequently ephemeral, but always actually spatially present. Minimalist sculptures are defined by intelligent, sensitive combinations of lines, surfaces, and volumes. With the simplest of means, his installations of acrylic yarns evoke remarkably illusionistic forms related exclusively to the classical canon of form-variants in Minimal Art. The acrylic yarns stretched between the architectural elements of ceiling, floor, walls, and corners play with the

sie dem, was wir bei klarem Wetter am Nachthimmel beobachten können, sehr ähnlich sehen, andererseits definitiv gänzlich von dem unterscheiden, was wir mit den beschränkten Mitteln unserer Mobilität und Zeit jemals werden sehen können. Dies schließt selbstverständlich aus, diese Ansichten als Repräsentationen einer Präsenz zu fassen. Die *Night Skies* sind Algorithmen und Befehlen geschuldet. Radikal gesprochen verabschiedet Angela Bulloch nicht nur ein spezifisches Repräsentationsdenken, sondern zugleich inauguriert sie eine Kunst, deren bestechende Ästhetik auf Algorithmen gründet, die der Software einer vorgängigen Hardware geschuldet sind.

Electric Wire Drawings

Analog zu den *Night Skies* als zweidimensionale Bilder mehrdimensionaler Räume des unendlichen Weltalls, bezeichnen die *Electric Wire Drawings* von Angela Bulloch zwar reale Körper, jedoch sind diese nicht als Masse im Ausstellungsraum vorhanden, sondern werden nur mittels ihrer Umrisslinien vorgestellt. Die *Electric Wire Drawings* nehmen mittelbaren und unmittelbaren Bezug auf die Arbeiten von Fred Sandback. Unmittelbaren, indem sie einzelne Arbeiten seines Oeuvres appropriieren und mittelbaren, indem sie andere Formen in abgewandelter Technik umsetzen. Das, was bei Fred Sandback einfache Garne oder Schnüre waren, transformiert Angela Bulloch in leuchtende Drähte. Das, was bei Sandback einfache und klare, mit möglichst wenig Bedeutung aufgeladene Objekte waren, wird bei Angela Bulloch zu Körpern, die einerseits durch die Referenz auf die Minimal Art Bedeutung erlangen und andererseits durch ihre Formenwahl (Platonische Körper) immer schon kulturgeschichtlich überdeterminiert sind.
Der US-amerikanische Künstler Fred Sandback (1943–2003) zählt zu den zentralen Figuren im Umfeld der Minimal Art. Seine häufig ephemeren, aber doch immer real im Raum präsenten, minimalistischen Skulpturen definieren sich durch eine intelligente und sensible Kombinatorik aus Linien, Flächen und Volumen. Mit einfachsten Mitteln rufen seine aus zum Teil vielfarbigen Acrylgarnen entwickelten Installationen bemerkenswert illusionistische Formen hervor, welche sich ausschließlich auf den klassischen Kanon der Formvarianten der Minimal Art beziehen. Die zwischen den architektonischen Grundelementen wie Decke, Boden, Wänden und Ecken gespannten Acrylgarne spielen mit den kognitiven und sinnlichen Fähigkeiten des Betrachters, welcher aus unterschiedlichen Standpunkten jeweils unterschiedliche Perspektiven auf das Kunstobjekt entwickelt. Form, Material und Raum sind so zueinander in Beziehung gesetzt, dass die interdependenten Elemente als ein komplexes Ganzes in Erscheinung treten. Zugleich überschreiten sie die Ambivalenz zwischen Außenraum und Innenraum, Subjekt und Objekt sowie Kunstobjekt und Architektur (Ausstellungsraum).

viewer's cognitive and sensory perceptions, each of which generate different perspectives on the art object from different viewpoints. Form, material, and space are put into relation with one another in such a way that the interdependent elements seem to become a complex whole. At the same time, these elements transcend the ambivalence between outside and inside, between subject and object, and between art object and architecture (exhibition space).

Bulloch appropriates Sandback's principle by reactivating two of his early works: *Untitled* (1967) and *Untitled* (1981), respectively a long rectangle lying on the floor (close to the dimensions of a two-by-four plank) and a line connecting the wall to the floor. She transforms Sandback's threads into illuminated wires, which not only bodilessly represent a solid body, as in Sandback's works, but which also radiate into the room, illuminating it with a glow. Following this principle, five other objects are represented by illuminated wires: a hexahedron (cube), a tetrahedron, an octahedron, a dodecahedron, and an icosahedron. These are the five convex polyhedra that Plato describes in the *Timaios* known as the Platonic solids (regular solids with many plane surfaces and corners at their outer extremities). Their surfaces are all congruent, regular polygons, each meeting in a corner and each with the same number of surfaces. Using these models, Plato wanted to explain the elements and their properties, or more precisely, the microcosm. In the early modern age, the astronomer and mathematician, Johannes Kepler, in his work *Mysterium cosmographicum* (1596), related these five Platonic solids to the heliocentric world view. By nesting the solids, Kepler tried to convey an idea of the spheres of the individual planets. The conception developed here of an order of the planets created by God on the basis of regular solids is one of Kepler's attempts to comprehend the cosmos as an harmonious whole. Later on, after observing the elliptical orbits of the planets, Kepler repeatedly made new attempts to relate this observation to harmonious, mathematical concepts. Bulloch's introduction of the Platonic solids into her work is a reference to these last scientific attempts to construe space as not only mathematically definable, but also as a representable structure under the premise in art history of a *philosophy of symbolic forms,* which understands all products of intellectual culture, such as language, myth, religion, art, and science, as an organic whole, and which simultaneously draws on all of the knowledge and methods of the individual disciplines for the sake of interpretation. When considered as a work of art at the beginning of the twenty-first century, the Platonic solids embody what is commonly described as a crisis of representation.

Within the context of this thought, Bulloch's *Electric Wire Drawings* paradigmatically embody the linear perspective discussed as a condition of art ever since art's invention, and simultaneously reflect this technique's illusionistic and mathematical construction.

Angela Bulloch unterzieht das Sandbacksche Prinzip einer Geste der Appropriation, indem sie zwei seiner wirkmächtigsten Arbeiten reaktiviert: *Untitled* von 1967 und *Untitled* von 1981. Ein langes am Boden liegendes Rechteck und eine horizontal verlaufende Linie, die Wand und Boden miteinander verbindet. Die Fäden von Sandback transformiert sie in selbstleuchtende Kabel, welche nicht nur, wie bei Sandback, körperlos einen Körper vorstellen, sondern in den Raum hinausstrahlen, ihn erleuchten und beleuchten. Weiterhin werden nach diesem Prinzip fünf weitere Körper durch die leuchtenden Kabel vorgestellt. Ein Hexaeder (Würfel), ein Tetraeder, ein Oktaeder, ein Dodekaeder und ein Ikosaeder. Es sind die fünf Platonischen Körper, die Platon im *Timaios* beschrieben hat. Als Platonische Körper werden diejenigen konvexen Polyeder (Vielflächner mit Ecken nach außen) bezeichnet, bei denen alle Flächen kongruente regelmäßige Vielecke sind und in einer Ecke jeweils gleich viele Flächen zusammentreffen. Platon hatte mit diesen Modellen die Elemente und deren Eigenschaften, genauer den Mikrokosmos, begreiflich machen wollen. In der Frühen Neuzeit verband der Astronom und Mathematiker Johannes Kepler in seinem Werk *Mysterium cosmographicum* (1596) diese fünf Platonischen Körper mit dem heliozentrischen Weltbild. Durch das Ineinanderschachteln der Körper versuchte Kepler die Sphären der einzelnen Planeten vorzustellen: Das hier entwickelte Verständnis einer von Gott geschaffenen, auf regelmäßigen Körpern beruhenden Planetenordnung ist einer der Versuche Keplers, das Weltall als harmonisches Ganzes zu begreifen. Später, nach der Beobachtung der elliptischen Planetenbahnen, unternahm Kepler immer neue Versuche, diese Beobachtung mit harmonischen, mathematischen Verhältnissen in Verbindung zu bringen. Insofern ist die Einführung der platonischen Formen durch Angela Bulloch ein Hinweis auf die letzten Versuche der Wissenschaft, den Raum als ein sowohl mathematisch beschreibbares als auch darstellbares Gefüge zu konstruieren. Im Sinne der kunsthistorischen Prämisse einer *Philosophie der symbolischen Formen,* alle Erzeugnisse der geistigen Kultur wie Sprache, Mythos, Religion, Kunst und Wissenschaft als ein organisches Ganzes zu betrachten und gleichzeitig zur Deutung alle Erkenntnisse und Methoden der Einzeldisziplinen heranzuziehen, verkörpern die platonischen Formen als Kunstobjekt zu Beginn des 21. Jahrhunderts das, was landläufig oft als Krise der Repräsentation beschrieben wurde.

Im Kontext dieser Überlegung verkörpern die *Electric Wire Drawings* von Angela Bulloch paradigmatisch die seit Erfindung der Kunst in Rede stehende Zentralperspektive als Bedingung der Kunst und reflektieren gleichzeitig ihre illusionistische und mathematische Konstruktion. Die erlernten Fähigkeiten des zeitgenössischen Betrachters hinsichtlich der räumlichen Wahrnehmung von Kunst werden durch die Wahrnehmungsanforderungen der leuchtenden Figuren produktiv erweitert. Simultan wird der Betrachter auf seine An-

The acquired abilities of the contemporary viewer with regard to the spatial perception of art are productively extended by the requirements of perception of the illuminated figures. Simultaneously, the viewer is thrown back upon his presence as a mere body, and as a context-bound contemplator of art as defined within cultural history. This double bind composed of physis and cultural convention thus not only reflects the status of the art object within the context of its institutional placement, but also always already reflects the corresponding figure of reflection of the viewer and the viewer's status as subject. Art history from the Renaissance to the present is thus consummated as a history not so much of the objects of interest within the discipline, but as a history of the subject's genesis mirrored in the figure of the viewer.

Whereas Minimal Art, with the reduction of form to simple solids, strove for a renewed theatricalization of the arts, Sandback's work initiated the relatively early and pioneering turn away from individual sculptural volumes toward an essentially extended conception of sculpture. Bulloch's works leave the classical status of art objects as "things in themselves" behind and act as a broad interface between the viewer, the art object, and the viewer's (institutional) surroundings—always already with the other potential viewers who populate this architecture that has been culturally determined as a white cube. Constructed of fine lines which leave space to move around the installations and through the room occupied by it, the works enable the viewers to perceive them while in motion. Still rooted in the idea of sculpture, but significantly expanded in the manner of installation art, the works undertake a redefinition of the fixed concepts of art. Whereas Sandback can be regarded as one of the most important intellectual consolidators of Minimal Art who prepared the ground simultaneously for installation art and institutional critique, Bulloch's works are not conceived in the sense of a consolidation. Rather, her works are the description of a crisis of space, of the idea of space, and perhaps even of installation art and its role within the exhibition space.

Translated from the German by Michael Eldred, artefact text & translation, Cologne

wesenheit als bloßer Körper und kulturhistorisch definierte wie kontextgebundene Figur des Kunstbetrachters zurückgeworfen. Dieses „Double bind" aus Physis und kultureller Konvention reflektiert somit nicht nur den Status des Kunstobjekts im Kontext seiner institutionellen Aufstellung, sondern immer auch die entsprechende Reflexionsfigur des Betrachters und seinen Subjektstatus. Damit vollendet sich die Kunstgeschichte gleichsam von der Renaissance bis zur Gegenwart als eine Geschichte weniger der Objekte ihres disziplinären Interesses, als einer Geschichte der Subjektwerdung, die in der Figur des Betrachters gespiegelt ist.

Während die Minimal Art mit der Reduktion auf simple Körper um eine Retheatralisierung der Künste bemüht war, hat die Arbeit von Fred Sandback als zusätzliche Komponente die relativ frühe und bahnbrechende Abkehr von einzelnen skulpturalen Volumina hin zu einem wesentlich erweiterten Skulpturbegriff eingeleitet. Angela Bullochs Arbeiten lassen den klassischen Status der Kunstobjekte als „Ding an sich" hinter sich und erweitern sich zu einer weitläufigen Schnittstelle zwischen Betrachter, Kunstobjekt und seiner (institutionellen) Umgebung – aber immer auch der potenziell anderen Betrachter, welche diese kulturell als Galerieraum überdeterminierte Architektur bevölkern. Konstruiert aus feinen Linien, die entsprechend Platz lassen, um sich durch den installativ besetzten Raum hindurch und um ihn herum zu bewegen, ermöglichen die Arbeiten eine Wahrnehmung in Bewegung. Immer noch geerdet im Begriff der Skulptur und doch bereits wesentlich erweitert hin zu einer konstruktiven Form der Installationskunst, beschreiten die Arbeiten eine Neudefinition fixierter Begrifflichkeiten der Kunst. Während Fred Sandback als einer der wichtigsten intellektuellen Konsolidierer der Minimal Art gesehen werden kann, der gleichzeitig der Installationskunst und der institutionskritischen Auseinandersetzung mit dem White Cube den Boden bereitet hat, verstehen sich die Arbeiten von Angela Bulloch gerade nicht im Sinne einer Konsolidierung. Vielmehr sind ihre Arbeiten die Beschreibung einer Krise des Raums, der Raumvorstellung und vielleicht sogar der Installationskunst und deren Zuständigkeit in Bezug auf den Ausstellungsraum.

Ausgestellte Arbeiten
Works Exhibited

1
***Electric Wire Drawing: P. S. Tetrahedron (floor, with points)*, 2008**
Drahtrahmen, weißes Elektrolumineszenz-Kabel, AC/DC-Wandler / Wire frame, white electroluminescent wire, inverter
170 × 182 × 211 cm
Courtesy Galerie Eva Presenhuber, Zürich

2
***Electric Wire Drawing: P. S. Hexahedron (floor, with points)*, 2008**
Drahtrahmen, weißes Elektrolumineszenz-Kabel, AC/DC-Wandler / Wire frame, white electroluminescent wire, inverter
163 × 200 × 229 cm
Courtesy Esther Schipper, Berlin

3
***Electric Wire Drawing: P. S. Octahedron (floor, with points)*, 2008**
Drahtrahmen, weißes Elektrolumineszenz-Kabel, AC/DC-Wandler / Wire frame, white electroluminescent wire, inverter
159 × 225 × 225 cm
Courtesy Cristina Guerra Contemporary Art, Lissabon

4
***Electric Wire Drawing: P. S. Dodecahedron (hanging, no points)*, 2008**
Drahtrahmen, weißes Elektrolumineszenz-Kabel, AC/DC-Wandler / Wire frame, white electroluminescent wire, inverter
111 × 132 × 140 cm
Courtesy Galerie Eva Presenhuber, Zürich

5
***Electric Wire Drawing: P. S. Icosahedron (hanging, no points)*, 2008**
Drahtrahmen, weißes Elektrolumineszenz-Kabel, AC/DC-Wandler / Wire frame, white electroluminescent wire, inverter
77 × 84 × 98 cm
Courtesy Esther Schipper, Berlin

6
***Last Year Again. 2008*, 2008**
Synchronisierte Doppelprojektion, digitale Videoanimation, 2 Mac minis, Videoprojektoren / Synchronised double projection, digital video animation, 2 Mac minis, video projectors
3´ 06˝
Courtesy Esther Schipper, Berlin

7
***Night Sky: Gleise 581C.12*, 2008**
LED-Installation, Neopren, Pollopasspiegel, Alu-Profile / LED-installation, Neoprene, Pollopas mirror, aluminium profile
Boden / floor 266,4 × 199,8 × 45 cm,
Wand / wall 266,4 × 199,8 × 20 cm
Courtesy Galería Helga de Alvear, Madrid

8
***Night Sky: Mars from Venus.12*, 2007**
LED-Installation, Neopren, Pollopasspiegel, Alu-Profile / LED-installation, Neoprene, Pollopas mirror, aluminium profile
Boden / floor 266,4 × 199,8 × 45 cm,
Wand / wall 266,4 × 199,8 × 20 cm
Courtesy Galerie Micheline Szwajcer, Antwerpen

9
***Night Sky: E.T. from Mercury.12*, 2007**
LED-Installation, Neopren, Pollopasspiegel, Alu-Profile / LED-installation, Neoprene, Pollopas mirror, aluminium profile
Boden / floor 266,4 × 199,8 × 45 cm,
Wand / wall 266,4 × 199,8 × 20 cm
Courtesy Esther Schipper, Berlin

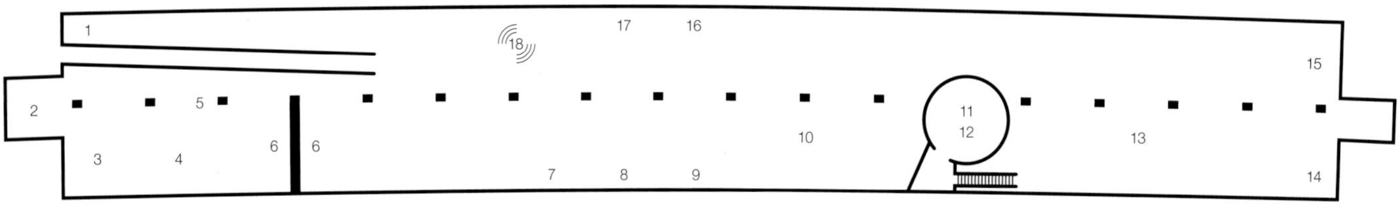

10
***Möbius Night Sky Model: Mark I*, 2003**
Stahl, Bitumen, LED-Installation, dmx Steuerung, Kabel und hölzerner Sockel / Steel, bitumen, LED-installation, dmx controller, cables and wooden plinth
71 × 70 × 130 cm
Collezione Nunzia e Vittorio Gaddi, Lucca

11
***Fly Over Genoa*, 1992/2008**
Digital überarbeitetes Hi8 Video, Mac mini und Videoprojektor / Digitally remastered Hi8 video, Mac mini and video projector
7´46˝
Courtesy die Künstlerin / the artist

12
***Flyover Genoa*, 1992/2008**
Digital überarbeitetes Hi8 Video, Mac mini und Videoprojektor / Digitally remastered Hi8 video, Mac mini and video projector
16´03˝
Courtesy die Künstlerin / the artist

13
***Progression of 8 Perverted Pixels*, 2008**
7 dmx Module, 1 black box Modul, dmx Steuerung und Kabel / 7 Plexiglas dmx modules, 1 Plexiglas black box module, dmx controller and cables
von / from 52 × 52 × 52 cm bis / to 62 × 70 × 62 cm
Courtesy Galerie Micheline Szwajcer, Antwerpen & Esther Schipper, Berlin

14
***Electric Wire Drawing: V Point (lime, wall to wall to wall)*, 2008**
Stahlpyramide, neongrünes Elektrolumineszenz-Kabel, AC/DC-Wandler / Steel pyramid, lime-green electroluminescent wire, inverter
176 × 348 × 433 cm
Courtesy die Künstlerin / the artist

15
***Z Point*, 2001/2004**
48 Plastik dmx Module, dmx Steuerung, Kabel, Audioanlage, Soundtrack von David Grubbs / 48 plastic dmx modules, dmx controller, cables, sound equipment, soundtrack by David Grubbs
300 × 400 × 50 cm
8´14˝
Städtische Galerie im Lenbachhaus und Kunstbau, München
Dauerleihgabe aus Privatbesitz / Permanent loan from private collector

16
***Electric Wire Drawing: F. S. Untitled (cyan, wall to floor)*, 2008**
Blaugrünes Elektrolumineszenz-Kabel, AC/DC-Wandler / Cyan electroluminescent wire, inverter
279 × 31 × 10 cm
Courtesy Esther Schipper, Berlin

17
***Electric Wire Drawing: F. S. Untitled (red, floor)*, 2008**
Stahl, rotes Elektrolumineszenz-Kabel, AC/DC-Wandler / Steel tubes, red electroluminescent wire, inverter
4,1 × 11,4 × 365,8 cm
Courtesy Esther Schipper, Berlin

18
Michael Iber & Angela Bulloch
***Z Phrase*, 2008**
24-Kanal Audioinstallation / 24-channel sound installation
9´13˝
Courtesy Esther Schipper, Berlin

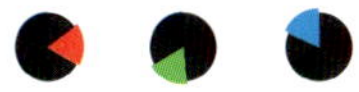

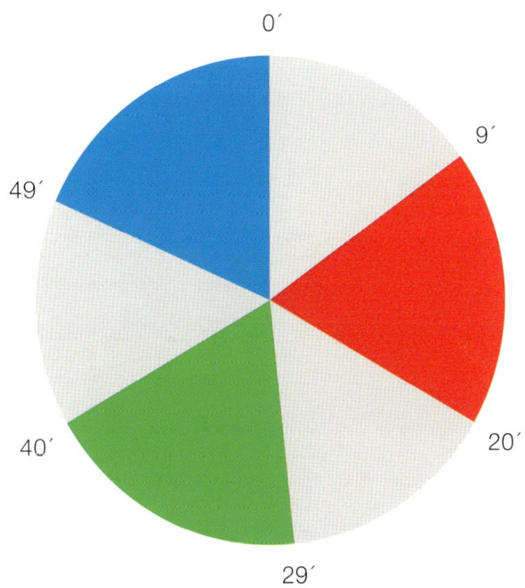

Z Point und *Z Phrase* wird drei Mal in der Stunde abwechselnd abgespielt.
Z Point and *Z Phrase* play alternately three times an hour.

Biographie
Biography

Angela Bulloch

1966	geboren / born in Rainy River, Kanada / Canada
1977	Umzug nach England / moved to England
1985–1988	Goldsmiths College, University of London, B.A. (Hons.) Fine Art
1999	Umzug nach Berlin, wo sie derzeit lebt und arbeitet / moved to Berlin, where she currently lives and works

Jüngste Einzelausstellungen / recent solo exhibitions

Jahr / year	Termin / dates	Ausstellungstitel / exhibition title	Ausstellungsort / exhibition venue	Ausgestellte Arbeiten / works exhibited
2008	15.3.–19.4.2008	*V*	Galerie Eva Presenhuber, Zürich	
2008	16.2.–18.5.2008	*The Space That Time Forgot*	Städtische Galerie im Lenbachhaus und Kunstbau, München (Kat. / cat.)	*Electric Wire Drawing: P. S. Tetrahedron (floor, with points); Electric Wire Drawing: P. S. Hexahedron (floor, with points); Electric Wire Drawing: P. S. Octahedron (floor, with points); Electric Wire Drawing: P. S. Dodecahedron (hanging, no points); Electric Wire Drawing: P. S. Icosahedron (hanging, no points); Last Year Again. 2008; Night Sky: Gleise 581C.12; Night Sky: Mars from Venus.12; Night Sky: E.T. from Mercury.12; Möbius Night Sky Model: Mark I; Fly Over Genoa; Flyover Genoa; Progression of 8 Perverted Pixels; Electric Wire Drawing: V Point (lime, wall to wall to wall); Z Point; Electric Wire Drawing: F. S. Untitled (cyan, wall to floor); Electric Wire Drawing: F. S. Untitled (red, floor); Z Phrase*
2007	11.5.–24.5.2007	*Repeat Refrain*	Ara Pacis, Roma	*Repeat Refrain*
2007	28.4.–23.6.2007	*Are you coming or going, around?*	Esther Schipper, Berlin	*Last Year.2007; Night Sky Prints E.T. From Mercury.12; Night Sky E.T. From Mercury.12; Smoke Spheres 2:4*
2006	24.6.–4.9.2006	*Angela Bulloch*	The Power Plant, Toronto (Kat. / cat.)	*RGB Spheres I; II; III; IV; Rock 'n' Roll Star: Byrds & Smith; The Power of 8; Frank: Group of Seven (One absent friend); RGB Disks; Gang of Four:16: Print No. 16*
2006	12.1.–7.5.2006	*Angela Bulloch*	De Pont, Tilburg (Kat. / cat.)	*RGB Spheres I; II; III; IV; Rock 'n' Roll Star: Byrds & Smith;* Pixel Wall Paintings – *21st Century Kanazawa, Lanzarote, Beijing; The Power of 8; Daniel: Group of Seven (One absent friend); Macro World: One Hour³ and Canned.; 6 Chains: Permutation B; RGB Disks; Construction Table House: Tourmaline; Gang of Four:16: Print No. 16; Flexible Single; Rules Series* edition
2005	21.12.2005–21.1.2006	*Angela Bulloch: Yuko Gothic Grid*	Galerie Micheline Szwajcer, Antwerpen	*Yuko: Group of Seven (One absent friend)*
2005	11.10.–18.12.2005	*Angela Bulloch*	Modern Art, Oxford (Kat. / cat.)	*Z Point; Solaris 1993; Chain A 4:1:20:4; Chain A 3:1:20:4; Chain A 2:1:20:4; Chain A 1:1:20:4; Fundamental Discord: 16; RGB Spheres I; II; III; IV; Rock 'n' Roll Star: Byrds & Smith; Utility Model Law; The King of Comedy; Construction Table House: Tourmaline; Construction Table Airplane: White; Flexible Single Lobster: Blue; Repressed Room; Rules Series* edition
2005	15.9.–13.11.2005	*Angela Bulloch: To the Power of 4.*	Secession, Wien (Kat. / cat.)	*Daniel / Frank / Yuko: Group of Seven (One absent friend); Video Sound System; various Flexible Singles*
2005	2.9.–16.10.2005	*Preis der Nationalgalerie für Junge Kunst 2005*	Hamburger Bahnhof – Museum für Gegenwart, Berlin (Kat. / cat.)	*The Disenchanted Forest x 1001; Posters-Goth Girl in a Dynamic Grid; Luther's Studio as seen through one of his works; Rastered Eye; Number 400; Rules Series: 1001 Plate; Tree Numbering Guide*

Jüngste Gruppenausstellungen / recent group exhibitions

Jahr / year	Termin / dates	Ausstellungstitel / exhibition title	Ausstellungsort / exhibition venue	Ausgestellte Arbeiten / works exhibited
2008	23.10.2008–1.2009	*Anyspacewhatsoever*	Solomon R. Guggenheim Museum, New York	
2008	6.3.–18.5.2008	*Martian Museum of Terrestrial Art*	Barbican Art Gallery, London	*Devil Bag*
2008	5.3.–29.6.2008	*Kunstmaschinen. Maschinenkunst*	Museum Tinguely, Basel (Kat./cat.)	*Blue Horizons*
2008	5.3.–5.4.2008	*TYPED*	Sadie Coles, London	*Smuggling*
2008	25.4.–13.7.2008	*Turner Prize: A Retrospective*	Mori Art Museum, Tokyo (Kat./cat.)	
2008	26.2.–12.5.2008	*Color Chart*	Museum of Modern Art, New York	*Standart Universal 256: Cyan*
2007	18.10.2007–31.1.2008	*Jubilee*	Eva Presenhuber's house, Vnà, Unterengadin	*WNBA Fan Code of Conduct*
2007	18.10.2007–27.1.2008	*Kunstmaschinen. Maschinenkunst*	Schirn Kunsthalle, Frankfurt/M. (Kat./cat.)	*Blue Horizons*
2007	2.10.2007–6.1.2008	*Turner Prize: A Retrospective*	Tate Britain, London (Kat./cat.)	
2007	8.9.–16.12.2007	*BODYPOLITCX*	Witte de With, Rotterdam (Kat./cat.)	*Baby-Doll Saloon*
2007	8.9.–4.11.2007	*CUESTIÓN XERACIONAL*	Centro Galego de Arte Contemporánea, Santiago de Compostela (Kat./cat.)	*Gang Of Four Blue: Print No. 11*
2007	7.9.–16.12.2007	*Ensemble*	Institute of Contemporary Art, University of Pennsylvania, Philadelphia (Kat./cat.)	*A 'B'*
2007	12.8.–7.10.2007	*COSMIC DREAMS*	La Kunsthalle, Centro Cultural Andratx, Andratx (Kat./cat.)	*Night Sky Prints E.T. From Mercury.12*
2007	14.2.– 29.4.2007	*Klang im Bild*	Stiftung Opelvillen, Rüsselsheim	*Rock 'n' Roll Star: Byrds & Smith*
2006	25.11.2006–28.1.2007	*In the Darkest Hours there may be Light: Works from Damien Hirst's murderme collection*	Serpentine Gallery London (Kat./cat.)	*Chain A 3:1:42:6; Random Reminder*
2006	22.10.2006–15.4.2007	*Bewegung im Quadrat*	Museum Ritter, Waldenbuch	*6 Chains: Permutation B*
2006	30.7.–24.9.2006	*Abstract art now – strictly geometrical*	Wilhelm Hack Museum, Ludwigshafen	*Power of 8*
2006	30.6.–22.10.2006	*Helga de Alvear – Concepts For a Collection*	Centro Cultural de Belém	*The Disenchanted Forest x 1001*
2006	27.1.–26.3.2006	*Satellite of Love*	Witte de With, Rotterdam	*Blow Up_T.V.*
2006	1.3.-14.5.2006	*Tate Triennial*	Tate Britain, London (Kat./cat.)	*The Disenchanted Forest x 1001*
2006	20.1.–18.3.2006	*Backdrop*	Bloomberg Space, London	*Chain A 2:1:20:4*

Architekturbezogene Kunstwerke / recent public commissions

2008	Public Art Project Zhang Jiang, Shanghai, China	
2007	Anyang Public Art Project 2007, Anyang, Korea	*Floral Tribute*
2006	Art Public Contemporain, Institut de Puericulture et de Perinatalogie, Paris, France	*Incubate Lactate Perambulate*
2005	Dior, Shinsaibashi, Osaka, Japan	*Anroidika Descending the Staircase.*

Eine umfangreichere Biographie finden Sie in / For a more comprehensive biography, see:
Prime Numbers Angela Bulloch. Secession, Wien; Modern Art Oxford, Oxford; De Pont, Tilburg; The Power Plant, Toronto. Verlag der Buchhandlung Walther König, Köln, 2006.

Bibliographie
Bibliography

Monographien / Monographs

2006
- *Prime Numbers Angela Bulloch.* Secession, Wien; Modern Art Oxford, Oxford; De Pont, Tilburg; The Power Plant, Toronto. Verlag der Buchhandlung Walther König, Köln.

2005
- *Angela Bulloch.* Secession, Wien. (Kat./cat.)

2002
- *Materials 01.* The Institute of Visual Culture, Cambridge. (Kat./cat.)

2001
- *Pixel Book. Angela Bulloch,* hrsg. von / edited by Beatrix Ruf. Kunsthaus Glarus, Glarus. (Kat./cat.)

2000
- *Rule Book. Angela Bulloch,* hrsg. von / edited by Stefan Kalmár. Book Works, London.
- *Documentary Notes. The Thames Archive Project: Henry Bond, Angela Bulloch.* Public Art Develoment Trust, London.

1998
- *Satellite. Angela Bulloch,* hrsg. von / edited by Stefan Kalmár & Angela Bulloch. Museum für Gegenwartskunst, Zürich; Le Consortium, Dijon. (Kat./cat.)

1994
- *Angela Bulloch.* CCC, Tours; FRAC Languedoc-Roussillon, Montpellier; Kunstverein, Hamburg. (Kat./cat.)
- *We Are Medi(eval). Angela Bulloch & Liam Gillick.* Portikus, Frankfurt/Main.

1990
- *Angela Bulloch,* hrsg. von / edited by Eric Troncy. APAC Centre d'Art Contemporain, Nevers.

Publikationen / Publications

2007
- *Anyang Public Art Project 2007.* Anyang Public Art Project, Anyang. (Kat./cat.)
- *Kunstmaschinen Maschinenkunst.* Schirn Kunsthalle, Frankfurt/Main. (Kat./cat.)
- *X Cuestión Xeracional.* Centro Galego de Arte Contemporánea, Santiago de Compostela. (Kat./cat.)

2006
- *Abstract Art Now. Strictly Geometrical?* Wilhelm-Hack-Museum, Ludwigshafen. (Kat./cat.)
- *Anstoss Berlin – Kunst macht Welt / Art makes the world.* Haus am Waldsee, Berlin. (Kat./cat.)
- Liam Gillick, Proxemics. *Selected Writings (1988–2006).* JRP Ringier, Zürich & Les Presses du Réel, Dijon.
- *Helga de Alvear. Concepts for a collection.* Centro Cultural de Belém, Lisboa. (Kat./cat.)
- *In the darkest hour there may be light: Works from Damien Hirst's murderme collection.* Serpentine Gallery, London. (Kat./cat.)
- *L'art dans la ville – 9 œuvres contemporaines sur le parcours du tramway T3 des Maréchaux Sud.* Mairie de Paris, Paris.
- *Lichtkunst aus Kunstlicht.* ZKM, Karlsruhe. Hatje Cantz Verlag, Ostfildern. (Kat./cat.)
- *Öffentliche Kunst, Kunst im öffentlichen Raum Niederösterreich, Band 8.* Springer Verlag, Wien & New York.
- *Tate Triennial 2006: New British Art.* Tate Britain, London. Tate Publishing, London. (Kat./cat.)
- *The Whitechapel Auction. Defining The Contemporary.* Sotheby's, London. (Kat./cat.)

2005
- *Art at 30 St Mary Axe London.* Swiss Reinsurance Company, Zürich.
- *Bidibidobidiboo. Works from Collezione Sandretto Re Rebaudengo.* Fondazione Sandretto Re Rebaudengo, Turin. Skira, Milano. (Kat./cat.)
- *Brightness.* Modern Museum of Art, Dubrovnik. (Kat./cat.)
- *Collection 2.* Fondation pour l'Art Contemporain Claudine et Jean-Marc Salomon, Alex. (Kat./cat.)
- *Des Deux Côtés du Rhin.* K21, Düsseldorf; Museum Ludwig, Köln. Snoeck, Köln. (Kat./cat.)
- *EN/OF 001–030.* Museum Kurhaus Kleve, Kleve. Revolver, Frankfurt/Main. (Kat./cat.)
- EVN 95–05. MUMOK, Wien. (Kat./cat.)
- *Extreme Abstraction.* Albright-Knox Art Gallery, Buffalo. (Kat./cat.)
- *Form Specific.* Moderna Galerija Ljubljana, Ljubljana. (Kat./cat.)
- *Preis der Nationalgalerie für junge Kunst.* Hamburger Bahnhof – Museum für Gegenwart, Berlin. (Kat./cat.)
- *Secession – Jahresbericht 2005.* Wiener Secession, Wien.
- *The Suspended Moment.* CRAC Alsace, Altkirch. (Kat./cat.)

2004
- *100 Artists See God.* Independent Curators International, New York. (Kat./cat.)
- Nicolas Bourriaud, *Postproduction.* Les Presses du Réel (Collection Documents sur l'art), Dijon.
- Xavier Douroux, *Nouveaux commanditaires en Bourgogne.* Les Presses du Réel, Dijon.
- *Ein-Leuchten.* Museum der Moderne, Salzburg. (Kat./cat.)
- *Funny Cuts, Cartoons und Comics in der zeitgenössischen Kunst.* Staatsgalerie Stuttgart, Stuttgart. Kerber Verlag, Bielefeld. (Kat./cat.)
- *Playlist.* Palais de Tokyo, Paris. Éditions Cercle d'Art, Paris. (Kat./cat.)
- *Villette Numérique 2004.* Parc et Grand Hall La Villette, Paris. (Kat./cat.)
- *Yellow Pages,* hrsg. von / edited by 404 Team & John Armleder. Ecart Publications, Geneva & JRP Ringier, Zürich.

2003
- *Brightness: Works from the Thyssen-Bornemisza Contemporary Art Foundation.* Museum of Modern Art, Dubrovnik. (Kat./cat.)
- *Conceptualisms.* Akademie der Künste, Berlin. (Kat./cat.)
- *Coollustre – Weather everything – Dramatically Different. Trois expositions d'Éric Troncy.* Collection Lambert, Avignon. Les Presses du Réel, Dijon. (Kat./cat.)
- *Der Dritte Sektor.* Kunstverein Wolfsburg; Galerie für Zeitgenössische Kunst, Leipzig, 2 Bd./vols.. (Kat./cat.)
- *Dreams and Conflicts – The Dictatorship of the Viewer. 50th International Art Exhibition.* La Biennale di Venezia, Venezia. Marsilio, Venezia. (Kat./cat.)
- *Einbildung – Das Wahrnehmen in der Kunst.* Kunsthaus Graz, Graz. Verlag der Buchhandlung Walther König, Köln. (Kat./cat.)
- *Episode – Chromatic Sensation by HERA.* Hyundai Gallery, Seoul. (Kat./cat.)
- *For your pleasure: Cai Guo-Qiang, Chiho Aoshima, Angela Bulloch.* Berkeley Art Museum, Berkeley. (Kat./cat.)
- *Frieze Art Fair Yearbook 2003–4.* Frieze, London. (Kat./cat.)
- *No Ghost just a Shell.* Van Abbemuseum, Eindhoven; Institute of Visual Culture, Cambridge; Kunsthalle Zürich, Zürich. Verlag der Buchhandlung Walther König, Köln. (Kat./cat.)
- *Performative Installation.* Siemens Art Programm, München; Galerie für Zeitgenössische Kunst, Leipzig. Snoek, Köln. (Kat./cat.)
- Gilda Williams (Hg. / ed.), *Cream 3. Contemporary Art in Culture, Phaidon,* London.

2002
- *2002 Shanghai Biennale.* Shanghai Art Museum, Shanghai. (Kat./cat.)
- *Claude Monet ... up to digital Impressionism,* Fondation Beyeler, Basel. Prestel, München. (Kat./cat.)
- Colin Painter (Hg. / ed.), *Contemporary Art and the Home.* Berg, Oxford & New York.
- *EVN Sammlung. Ankäufe 2000–2002.* EVN AG, Wien.
- *Frequenzen [Hz] / Frequencies [Hz].* Schirn Kunsthalle, Frankfurt/Main. Hatje Cantz Verlag, Ostfildern. (Kat./cat.)
- *Kunst am Bau – Die Projekte des Bundes in Berlin.* Bundesministerium für Verkehr, Bau- und Wohnungswesen. Ernst Wasmuth Verlag, Tübingen & Berlin.
- *Once Again.* John Hansard Gallery, Southampton. (Kat./cat.)
- *Remix.* Tate Liverpool. Tate Publishing, London. (Kat./cat.)
- Burkhard Riemschneider & Uta Grosenick (Hg. / eds.), *Art Now – 137 Artists at the Rise of the New Millennium.* Taschen, Köln.
- *Sweet Nothing – (Liege)Stätten des sommerlichen Nichtstuns.* Kunsthaus Baselland, Muttenz. (Kat./cat.)

Touch – Relational Art from the 1990s to Now. San Francisco Art Institute, San Francisco. (Kat./cat.)

2001

6e Biennale de Lyon art contemporain: Connivence 2001, prélude à 2003. Biennale de Lyon, Lyon. (Kat./cat.)

Arbeit Essen Angst: ein Kunstprojekt der Stiftung Industriedenkmalpflege und Geschichtskultur. Kokerei Zollverein, Essen. (Kat./cat.)

Art > Music: Rock, Pop, Techno. Museum of Contemporary Art, Sydney. Hatje Cantz Verlag, Ostfildern. (Kat./cat.)

Uta Grosenick (Hg. / ed.), *Women Artists. Künstlerinnen im 20. und 21. Jahrhundert.* Taschen, Köln.

TIMEWAVE ZERO / The Politics of Ecstasy. Grazer Kunstverein, Graz. (Kat./cat.)

2000

Against Design. ICA, University of Pennsylvania, Philadelphia. (Kat./cat.)

Marius Babias & Florian Waldvogel, *Campus 2002: Bildungsbegriffe.* Kokerei Zollverein, Essen.

Dire AIDS. Arte nell'epoca dell'AIDS. Promotrice delle Belle Arti, Torino. Edizioni Charta, Milano. (Kat./cat.)

Ein | Räumen. Arbeiten im Museum. Hamburger Kunsthalle, Hamburg. Hatje Cantz Verlag, Ostfildern. (Kat./cat.)

Media_City Seoul 2000. Seoul Metropolitan Museum, Seoul. (Kat./cat.)

Présumés innocents. L'art contemporain et l'enfance. CAPC Musée d'Art Contemporain, Bordeaux. (Kat./cat.)

Sonic Boom – The Art of Sound. Hayward Gallery, London. Hayward Gallery Publishing, London. (Kat./cat.)

There is something you should know. Die EVN Sammlung im Belvedere. Österreichische Galerie im Belvedere, Wien. (Kat./cat.)

Version_2000. Centre pour l'image contemporaine Saint-Gervais, Geneva. (Kat./cat.)

1999

At Home with Art, hrsg. von / edited by Colin Painter. Tate Gallery, London. Hayward Gallery Publishing, London. (Kat./cat.)

EVN Sammlung. Ankäufe 1997/99. EVN, Maria Enzersdorf.

Mehdi Chouakri Berlin. Yearbook 1998/99. Mehdi Chouakri, Berlin.

Power: Angela Bulloch, Kendell Geers, Gregory Green, Fabrice Gygi. Galerie für Zeitgenössische Kunst, Leipzig. (Kat./cat.)

Qu'est-ce que l'art (aujourd'hui) / What is art (today). Beaux Arts magazine, numéro special, 15.12.1999

Burkhard Riemschneider & Uta Grosenick (Hg. / eds.), *Art at the Turn of the Millennium.* Taschen, Köln.

This is Yesterday / Esto es Ayer. Centro Andaluz de Arte Contemporáneo, Seville. (Kat./cat.)

1998

Crossings: Kunst zum Hören und Sehen. Kunsthalle Wien, Wien. Cantz Verlag, Ostfildern. (Kat./cat.)

Die Sammlung Paul Maenz. Neues Museum, Weimar. Cantz Verlag, Ostfildern. (Kat./cat.)

Dimensions Variable. Städtische Kunstsammlungen, Chemnitz; Institut Mathildenhöhe, Darmstadt. The British Council. London. (Kat./cat.)

Emotion: young British and American art from the Goetz Collection. Deichtorhallen, Hamburg. Cantz Verlag, Ostfildern. (Kat./cat.)

Here to stay, hrsg. von / edited by Susan May. Hayward Gallery, London. Hayward Gallery Publishing, London. (Kat./cat.)

Le Consortium – une experience de l'exposition. Compilation. Centre Georges Pompidou, Paris. Les Presses du Réel, Dijon. (Kat./cat.)

Mai 98: Positionen zeitgenössischer Kunst seit den 60er Jahren. Kunsthalle Köln, Köln. Oktagon, Köln. (Kat./cat.)

Root. Thurston Moore. LO Recordings, London.

1997

Art from the UK: Angela Bulloch, Willie Doherty, Tracey Emin, Douglas Gordon, Mona Hatoum, Abigail Lane, Sarah Lucas, Sam Taylor-Wood, Rachel Whiteread. Sammlung Goetz, München. Kunstverlag Invild Goetz, München. (Kat./cat.)

EVN Sammlung. Ankäufe 1995/96. Schloß Ottenstein. Ottenstein. EVN, Maria Enzersdorf. (Kat./cat.)

Flexible. Museum für Gegenwartskunst, Zürich. (Kat./cat.)

Kunst.Arbeit. Südwest LB, Mannheim. Cantz Verlag, Ostfildern.

life / live – A cena artistica no Reino Unido em 1996 novas aventuras. Centro de Exposições, Centro Cultural de Belem, 2 Bd./vols.. (Kat./cat.)

Performance Anxiety. Museum of Contemporary Art, Chicago. (Kat./cat.)

Private Face / Urban Space. A new generation of artists from Britain. Rethymnon Centre for Contemporary Art, Rethymnon. (Kat./cat.)

Brigitte Oetker & Christiane Schneider (Hg. / eds.), *[Realisation] Kunst in der Leipziger Messe.* Oktagon, Köln.

The Turner Prize 1997. Tate Gallery, London. Tate Gallery Publishing, London. (Kat./cat.)

Un Toit pour tout le monde. FRAC Languedoc-Roussillon, Montpellier. (Kat./cat.)

Virginia Button, *The Turner Prize.* Tate Gallery, London. Tate Publishing, London.

1996

Collezionismo a Torino, Le opere di sei collezionisti d'Arte Contemporenea. Castello di Rivoli – Museo d'Arte Contemporenea, Torino. Edizioni Charta, Milano. (Kat./cat.)

Gedraag je! / Behave! / Benimm Dich! / Tiens-toi bien! Stedelijk Museum Bureau Amsterdam, Amsterdam. (Kat./cat.)

Jahresgaben 1996. Kölnischer Kunstverein, Köln.

life / live – la scène artistique au Royaume Uni en 1996 de nouvelles aventures. Musée d'Art Moderne de la Ville de Paris, Paris, 2 Bd./vols.. (Kat./cat.)

nach weimar. Landesmuseum & Schlossmuseum, Weimar. Kunstsammlungen zu Weimar, Weimar. (Kat./cat.)

Projekte / Projects, Kunst in der Neuen Messe Leipzig, hrsg. von / edited by Christiane Schneider & Brigitte Oetker. Leipziger Messe, Leipzig.

Traffic. CAPC Musée d'Art Contemporain, Bordeaux. (Kat./cat.)

1995

Am Rande der Malerei. Kunsthalle Bern, Bern. (Kat./cat.)

Collection, fin XXe, 1983–1995, douze ans d'acquisition d'art contemporain en Poitou-Charentes. FRAC Poitou-Charentes, Angouleme. (Kat./cat.)

Corpus Delicti, London i 90'erne. Kunstforeningen, Copenhagen. (Kat./cat.)

Das Ende der Avantgarde – Kunst als Dienstleistung, Sammlung Schürmann. Kunsthalle der Hypo-Kulturstiftung, München. (Kat./cat.)

Klangskulpturen Augenmusik: Grenzgänge zwischen Musik und Plastik im 20. Jahrhundert. Ludwig Museum im Deutschherrenhaus, Städtische Museen, Koblenz. (Kat./cat.)

Space Odyssey. Eleni Koroneou Gallery, Athens. (Kat./cat.)

Temporary Translation(s) – Sammlung Schürmann. Kunst der Gegenwart und Fotografie. Deichtorhallen Hamburg, Hamburg. (Kat./cat.)

Territorio Italiano – progetto d'eternità per l'arte contemporanea. Giacinto di Pitrantonio, Milano.

1994

LOW HOTEL – Your Place in the world: Angela Bulloch, Meg Cranston, Sarah Seager, Thaddeus Strode. Art & Public, Geneva; 1301, Santa Monica. (Kat./cat.)

Wall to Wall. Serpentine Gallery, London; Southampton City Art Gallery, Southampton; Leeds City Art Gallery, Leeds. The Southbank Centre, London. (Kat./cat.)

1993

Documentario 2. Spazio Opos, Milano.

Marjorie Jacobsen, *Art & Business.* Thames & Hudson, London.

Jahresring 41: Jahrbuch für moderne Kunst. Verlag Silke Schreiber, München.

L'Endroit Ideal (Ideal Place). L'Ile du Roy, Centre d'art et Jardin, Val de Reuil.

Viennese Story. Wiener Secession, Wien. (Kat./cat.)

Nulle Part et Partout. Carte blanche à Ami Barak. Espace Paul Boyé, Sète. (Kat./cat.)

1992 *Angela Bulloch, Adam Chodzko, Matthew Collings, Jeremy Deller, Douglas Gordon, Gary Hume, Michael Landy, Jonathan Monk, Simon Patterson, Brendan Quick, Philip Riley, Caroline Russell, Giorgio Sadotti, Gavin Turk, Gillian Wearing – Instructions received by Liam Gillick.* Gió Marconi, Milano. (Kat./cat.)

Dirty Data – Sammlung Schürmann. Ludwig Forum für internationale Kunst, Köln. Wilhelm Schürmann, Aix-la-Chapelle. (Kat./cat.)

États specifiques. Musée des Beaux-Arts André Malraux, Le Havre. (Kat./cat.)

Il faut construire l'hacienda. CCC, Tours. (Kat./cat.)

Molteplici Culture. Convento de S. Egidio, Roma. Edizione Carte Segrete, Roma. (Kat./cat.)

ON – Henry Bond, Angela Bulloch, Liam Gillick, Graham Gussin, Markus Hansen. Interim Art, London. (Kat./cat.)

1991 *Art>Music. An exhibition looking at crossovers between visual art and music.* Museum of Contemporary Art, Sydney. (Kat./cat.)

Broken English. Serpentine Gallery, London. (Kat./cat.)

No Man's Time. Centre national d'art contemporain, Villa Arson, Nice. (Kat./cat.)

Plastic Fantastic Lover. BlumHelman Warehouse, New York. (Kat./cat.)

Wechselstrom. Sammlung Hauser und Wirth / Teil 2. Sammlung Hauser und Wirth in der Lokremise, St. Gallen. Oktagon, Köln. (Kat./cat.)

1990 *Bulloch, Hume, Landy.* Esther Schipper, Köln. (Kat./cat.)

Nachschub, hrsg. von / edited by SPEX Magazine. Köln Show, Köln. (Kat./cat.)

1989 Home Truths. Castello di Rivara, Turin. Edizioni Franz Paludetto, Torino. (Kat./cat.)

1988 *Freeze.* PLA Building, London. (Kat./cat.)

Essays, Rezensionen / Articles, Reviews

2008 Cornelia Gockel, „Wenn Pixel explodieren": *Süddeutsche Zeitung,* 15.02.2008.

Adina Popescu, „Angela Bulloch": *Artforum International,* January, 2008, 163.

Birgit Sonna, „Die digitale Wunderlampe": *Art – Das Kunstmagazin,* 2, 2008, 60–65.

2007 Lauretta Colonelli, „Luce lunare sull'Ara Pacis": *Corriere della Sera,* 05.05.2007, 17.

Zoë Gray, „Angela Bulloch interview": *Witte de With Season Programme Autumn/Winter 2007,* Witte de With, Rotterdam, 24–25.

„L'arte si colora di luce": *E Polis Roma,* 05.05.2007, 41.

Renata Mambelli, „Ara Pacis, l'arte sulla luna": *la Repubblica,* 05.05.2007, 3.

Anna Lisa Martella, „L'arte contemporanea accende la lampadina": *Il Messagero,* 05.05.2007, 47.

Chiara Pellegrini, „Tre luoghi della città illuminati dagli artisti con maggiore ‚energia'": *Libero,* 05.05.2007, 50.

Felix Prinz, „Welt aus den Angeln": *Texte zur Kunst,* XVII/67, September, 252–254.

Philippe Régnier, „L'art transforme la ville": *Le Journal des Arts,* 269, 29.11.2007, 12.

Marilisa Rizzitelli, „L'energia al servizio dell'arte": *Inside Italia,* 30.05.2007, 49.

Chiara Somajni, „Cadono Lune": *Ventiquattro,* 05.05.2007, 104–107.

„The Producers": *Artforum International,* XLVI/2, October, 352–359, 402.

t.ma., „Mecenatismo Illuminato": *il Riformista,* 05.05.2007, 5.

Sabrina Vedevotto, „Ecco un'arte molte ‚illuminante'": *il Giornale,* 10.05.2007, 49.

2006 Dominic van den Boogerd, „Pixels en regels": *Metropolis M,* 1, 14–15.

Sacha Bronwasser, „Prettig frivol licht van Angela Bulloch": *Volkskrant,* 09.02.2006.

Frits de Coninck, „Betoverende pixels": *museum tijdschrift,* XIX/2, March / April, 34–37.

Edo Dijksterhuis, „Specialistenblind": *Het Financieele Dagblad / FD Persoonlijk,* 11.02.2006, 18–19.

Martin Herbert, „1000 Words: Angela Bulloch": *Artforum International,* XLIV/8, 218–219.

Gerrit van den Hoven, „Angela Bulloch gaat door waar Judd stopt": *Brabants Dagblad,* 02.01.2006.

Tom Morton, „Definitely, Maybe": *frieze,* 98, April, 124–129.

Charlotte Mullins, „All the world's a studio ...": *The Independent on Sunday,* 26.02.2006, ABC, 6–9.

Adrian Searle, „Guff and Nonsense": *The Guardian,* 28.02.2006, G2, 18–20.

Kim Seung-duk, „What Difference Does It Make?": *Frog,* 3, Spring/Summer, 112–117.

Kim Seung-duk, „Power Interview, Angela Bulloch's Pixel Odessey": *Art in Culture,* 4, 102–107.

Janneke Wesseling, „Baarmoedersentsatie": *NRC Handelblad,* 20.01.2006, 20–21.

2005 John Bock, „Voll Bock": *Kultur Spiegel,* 9, September, 14–18.

Michael Bracewell, „Angela Bulloch": *Art Review,* December, 141.

Angela Bulloch, „Ascend and Descend into the Light": *Sleek Magazine,* Autumn, 96–101.

Paula Clark, „Angela Bulloch": *Oxfordshire Living,* November, 63.

Diedrich Diedrichsen, „Echos von Spiegelsounds in Headphones": *Texte zur Kunst,* Dezember, XV/50, 42–61.

Christopher Gray, „Seeing the point of Bulloch's work": *The Oxford Times,* 02.12.2005, 2.

Jörg Heiser, „Wie eine launische Cyborg-Katze": *Süddeutsche Zeitung,* 03./04.09.2005, 15.

Faye Hirsch, „Abstract Generations": *Art in America,* October, 123–129.

Ana Finel Honigman, „Suzanne Cotter": *Contemporary,* 77, 46–49.

Oliver Koerner von Gustorf, „Kopf über, Beine breit": *Welt am Sonntag,* 04.09.2005, 65.

Susanne Leeb, „Possession": *Texte zur Kunst,* Dezember, XV/50, 192–196.

Mariano Mayer, „Angela Bulloch": *Ae,* 36, January, 22–23.

Mariano Mayer, „Angela Bulloch": *Neo2,* 40, February, 68–69.

Markus Mittringer, „Vom Eigenleben der Pixel": *Der Standard,* 15.09.2005.

Helen Peacocke, „Basking in Light": *Oxford Times Weekend Supplement,* 14.10.2005.

Mónica Rebollar, „Angela Bulloch": *Lápiz,* 210–211, February–March, 189.

ripe., „Kunst": *Frankfurter Allgemeine Sontagszeitung,* 04.09.2005, 32.

Guillermo Solana, „Películas ocultas de Angela Bulloch": *El Cultural,* 27.01.2005, 26–27.

Maria Walsh, „The Light Fantastic": *Art Monthly,* 292, December / January, 1–6 (+ Cover).

2004 Angelika Affentranger-Kirchrath, „Abstraktion in Pixelform": *Neue Zürcher Zeitung,* 33, 10.02.2004, 48

Katja Blomberg, „Kunstrakete auf Traufhöhe": *Süddeutsche Zeitung,* 91, 20.04.2004, 13.

Alex Coles, „Angela Bulloch: A Storm in a Pixel": *Contemporary,* 65, 20–23.

"Einblick 51": *Tageszeitung,* Berliner Ausgabe, 23.06.2004, 28.

Hilary Stunda, „Aspen: A Cutting-Edge Museum and First-Rate Galleries": *Artnews,* January.

2003 Martin Behr, „Betrogen von den eigenen Augen. Grazer Kunsthaus eröffnet den Ausstellungsbetrieb: Einbildung – das Wahrnehmen in der Kunst": *Salzburger Nachrichten,* 25./26.10.2003

Michel Gauthier, „une banquette contre l'aura": *Vacarme,* 25, Autumn, 76–78.

Stewart Oksenhorn, „Not necessarily the news": *The Aspen Times,* 20./21.12.2003, B1, B7, B9.

Nina Schedlmayer, „Hinters Augenlicht geführt": *Profil,* 45, 03.11.2003, 117

Helmut Schödel, „Der eingebildete Kranke": *Süddeutsche Zeitung,* 21.10.2003.

2002 Michael Archer, „Angela Bulloch": *Camera Austria,* 80, 84–5.

Diana Baldon, „Institute of Visual Culture, Cambridge": *tema celeste,* 94, 112.

Pierre Yves Desaive, „Angela Bulloch: Pixels & Rules": *Flux News,* 29, September/ November 2002, 6.

Power Ekroth, „Konsthimmel över Berlin": *Res,* 10, 40–46.

Alex Farquharson, „Angela Bulloch": *Frieze,* 71, November / December, 97.

Harald Fricke, „Frequenzen (HZ)": *Artforum International,* 41, Oktober 2002, 164.

Justin Hoffmann, „Timewave Zero – The Politics of Ecstasy": *Kunstforum International,* CLVIII, März, 367–368.

Holger Liebs, „Dingsda": *Süddeutsche Zeitung,* 19.08.2002, 14.

Alexander Marzahn & Karin Gerig, „Die Kunst, es so zu sagen, als wär es nie gesagt gewesen": *Basler Zeitung,* 20.08.2002, 34.

John Miller, „United Colors of BBC": *Texte zur Kunst,* XII/47, September, 153–155.

Mai Thu Perret, „Angela Bulloch": *Frieze,* 65, March, 89.

Martin Prinzhorn, „Simulation der Simulation (und umgekehrt) / The Simulation of Simulation (and vice versa)": *Parkett,* 66, 50–59.

Juliane Rebentisch, „Angela Bullochs Digitale Reduktionen / Angela Bulloch's Digital Reductions": *Parkett,* 66, 20–35.

Andrew Wilson, „Maybe / Vielleicht": *Parkett,* 66, 36–49.

2001 Jennifer Allen, „Angela Bulloch": *Artforum International,* February, 161.

"Blurring the boundaries and escaping the cubes": *The Independent on Sunday,* 10.06.2001.

Mark Currah, „Angela Bulloch": *Time Out,* 20.06.2001, 58.

"Das Pixel in der Kunst": *Fridolin,* 13.09.2001, 13.

Martina Koch, „Stellwerk der Wahrnehmung": *St. Galler Tagblatt,* 13.09.2001, 23.

Claudia Kock Marti, „Auch ‚Z Point' für Beatrix Ruf": *Glarner Nachrichten,* 12.09.2001, 11.

„Kunterbunte Welt in der Farbbox": *Glarner Nachrichten,* 06.09.2001, 31.

Nina Möntmann, „Institutional Critique vs. Corporate Identity": *Texte zur Kunst,* XI/41, März, 177–182.

Jens Rönnau, „Ein | Räumen – Arbeiten im Museum": *Kunstforum International,* 1–3, 323–326.

Beatrix Ruf, „Angela Bulloch thematisiert Ordnungssysteme mit Fehlern": *Impuls,* 4, December / January, 12.

Marina Sorbello, „Angela Bulloch": *tema celeste,* 83, 102.

2000 Ronald Berg, „Das Pixel lügt": *Zitty,* 2, 66.

Christoph Blase, „Leuchte, Pixel, leuchte hell": *Frankfurter Allgemeine Zeitung,* 01.11.2000, B4.

Knut Ebeling, „Das Maß der Pixel": *Der Tagesspiegel,* 04.11.2000, 28.

Alison Green, „Angela Bulloch *Rule Book,* Janice Kerbel *15 Lombard St.,* Nils Norman *The Contemporary Picturesque": Art Monthly,* 242, December / January, 54.

William L. Hamilton, „New Art's Interior Motive: A Generation's Credo": *The New York Times,* 03.02.2000, 6–8.

Judith Jammers, „Pixel in der Dingwelt": *Berliner Zeitung,* 23.10.2000, 16.

Jonathan Jones, „A Home Of Your Own": *Frieze,* 50, January / February, 50–53.

„Kunst-Szenen. In Pixeln sehen – Angela Bulloch": *Neue Zürcher Zeitung,* 12.04.2000, 46.

Marion von Osten, „Es läuft alles nach Plan": *Texte zur Kunst,* X/39, September, 189–196.

Christian Röllin, „Art and Communication": *Winfo,* March, 17–20.

Birgit Sonna, „Verführung zur Kunst": *Lufthansa Magazin,* 4, 6–18.

Edward J. Sozanski, „Along the fuzzy boundary between design and art": *The Philadelphia Inquirer,* 13.2.2000, 11.

1999 Charlotte Abrahams, „Art to Die for": *The Guardian Weekend,* 09.10.1999, 68–70.

Roy Exley, „Angela Bulloch: Robert Prime Gallery. London, England": *ZingMagazine,* II, Winter, 183–185.

Sigrun Hellmich, „Hochexplosiver Stoff?": *Sächsische Zeitung,* 15./16.5.1999.

Hilary Jay, „Are we having fun yet? Wildest Dreams": *Inquirer Magazine,* 11.04.1999, 11–15 (+ Cover).

Manuel Bonik, „Kunsthauptstadt Berlin": *GQ,* 12, 60–68.

Ulrich Müller, „Von Phall zu Fall": *Zitty,* 13, 15.06.1999, 57.

Tim Sommer, „Wechselspiel von Macht und Unterordnung": *Leipziger Volkszeitung,* 19.04.1999.

1998 Sabine Altorfer, „Smarte Kunst ohne Berührungsängste": *Limmattaler Tagblatt,* 29.04.1998.

„Angela Bulloch im Museum für Gegenwartskunst": *Kunst-Bulletin,* Juni, 56.

„Angela Bulloch ‚Superstructure'": *Schweizer Monatshefte,* 05.05.1998.

Knut Ebeling, „Wie die Lichtorgel im Partykeller": *Der Tagesspiegel,* 04.07.1998, 28.

Fabienne Fulchéri, „Bouffée d'oxygène au Havre": *Technikart,* 21, 100.

Michael Hauffen, „Angela Bulloch – Superstructure": *Kunstforum International,* 141, Juli / September, 420.

Dora Imhof, „Zürich: Angela Bulloch": *Basler Zeitung,* 27.05.1998.

Ken Johnson, „Inglenook": *The New York Times,* 25.07.1998, E39.

Gisela Kuoni, „Eine große Bühne für die Kunst": *Bündner Zeitung,* 28.04.1998.

Karl-Iversen Lapp, „Sie tanzen, spielen, stellen Masken her": *Tagesanzeiger,* 30.04.1998.

Norman Lindner, „Installationen von Angela Bulloch in der Galerie Schipper & Krome": *Tageszeitung,* Berliner Ausgabe, 23.06.1998, 24.

Gerhard Mack, „Gelegenheit zum Gedächtnisverlust": *Cash,* 15.05.1998.

Susan May, „Angela Bulloch": *Here to Stay. Arts Council Collection Purchases of the 1990s.* Arts Council London, 1998, 22–23.

David Perreau, „Replay": *Omnibus,* 25, July, 13.

Martin Pesch, „Die Mauer muß wieder her – Gegen die kunsthistorische Okkupation von Popkultur und Musik": *Neue Bildende Kunst,* 5, Oktober / November, 23–31.

Brita Polzer, „Dorothea Lange / David Hockney / Angela Bulloch": *Annabelle,* 01.05.1998.

Hans Rudolf Reust, „Angela Bulloch": *Artforum International,* 37, December, 138.

„Superstructures with Satellites. Angela Bulloch im Museum für Gegenwartskunst Zürich": *Neue Zürcher Zeitung,* 28.04.1998.

Jan Winkelmann, „Angela Bulloch": *Artist,* 35, 4–7 (+ Cover).

1997 Daya Alberge, „Fluff? Hair? Must be showtime at the Tate again": *The Times,* 29.10.1997, 5.

Richard Alleyne, „Women Behaving Badly (also known as the Turner Prize)": *Daily Mail,* 18.06.1997, 13.

Denis Angus, „La grande famille": *Technikart,* June 1997.

„All aboard the singing doughnut (yes, it's Turner Prize time again)": *The Independent on Sunday,* 26.10.1997.

Louisa Buck, „Life, art and the Turner women": *Evening Standard,* 28.10.1997, 28–29.

Angela Bulloch, „Go shoot yourself": *The Independent Saturday Magazine,* 13.12.1997, 26–32.

David Bussel, „Angela Bulloch": *Frieze,* 35, June / August, 85.

Dalya Alberge, „All-women shortlist takes Turner by surprise": *The Times,* 18.06.1997, 5.

Rachel Campbell-Johnston, „The Turner pickles its sacred cows": *The Times,* 18.06.1997, 18.

Matthew Collings, „Girls win prizes": *The Independent,* 17.06.1997, 6–7.

Martin Coomer, „Electronic Art": *Time Out,* 05.11.1997, 48.

Martin Coomer, „Prized apart. Who will take the Turner this year?": *Time Out,* 05.11.1997, 48.

Emma Crichton-Miller, „Turner Prize": *Frank,* 41–42.

Barry Didcock, „Four in the frame for prize surprise": *The Scotsman,* 19.06.1997.

Richard Dorment, „'Ello, 'ello – what's going on here?": *The Daily Telegraph,* 05.11.1997, 22.

William Feaver, „Something in the air": *The Observer Review,* 02.11.1997, 10.

William Feaver, „The jury's still out, but where's the spice, girls?": *The Observer,* 22.06.1997.

„Female quartet on Turner shortlist": *Wrexham Evening Leader,* 01.10.1997.

Annie Flury, „Turner-up for art books as painters, sculptors (and men) passed over": *The Birmingham Post,* 01.10.1997.

Ian Gallagher, „Very nice, but is it art ladies?", *The Express,* 18.06.1997, 21.

Dan Glaister, „A woman's place – in the gallery": *The Guardian,* 18.08.1997, 3.

Dan Glaister, „Art of anger launches Turner Prize": *The Guardian,* 29.10.1997, 9.

Jonathan Glancey, „Why this woman is a work of art": *The Guardian,* 21.06.1997, 6.

Hugo Glendinning, „art that's a blast": *Telegraph Magazine,* 25.10.1997, 52–55.

Andrew Graham-Dixon, „The Female Gaze": *Vogue,* December, 235–239.

David Green, „Bringing It All Back Home": *Contemporary Visual Arts,* 16, 1997, 54–60.

Martin Hannan, „A dirty shirt, two tables, some old bones and a midget in a bath...it's Turner time again": *The Scotsman,* 18.06.1997.

Mark Harris, „Angela Bulloch at Robert Prime": *Art in America,* May, 135–136.

Mark Harris, „Immaterial Culture": *Art Monthly,* 206, May, 6–8.

Katharina Hegewisch, „Musterkollektion des Zeitgeistes: Kunst in der Leipziger Messe": *Frankfurter Allgemeine Magazin,* 27.06.1997, 36–41.

Mike Higgins, „New art, old issues": *The Independent (Eye on Wednesday),* 29.10.1997, 6.

Tim Hilton, „The gallery that likes to make a meal of its art": *The Independent on Sunday,* 02.11.1997, 7.

Margaret Hussey, „If you take a good look you might even like it": *The Express,* 29.10.1997.

Andrea Jahn, „Angela Bulloch: Sound Bank": *Springer,* III/3, 73.

Waldemar Januszczak, „Women on the verge ...": *The Sunday Times,* 02.11.1997, 12.

Waldemar Januszczak, „Women on top": *The Sunday Times,* 22.06.1997, 9.

Liz Jobey, „A rat race?": *The Guardian Weekend,* 04.10.1997, 28–37.

David Lee, „Britain's art world is in the grip of a Soviet-style elite. For socialist realism, read conceptualism. By the way, they're ripping off the taxpayer too": *New Statesman,* 04.07.1997, 40–41.

David Lister, „Serota defends all-woman Turner Prize shortlist": *The Independent,* 18.06.1997, 2.

David Lister, „Turner Prize set for all-women shortlist": *The Independent,* 16.06.1997.

Martin Love, „Look who's wearing the trousers now": *Evening Standard Magazine,* 21.11.1997, 16–17.

Dominic Lutyens, „Gallery Grrrls": *The Times,* 01.11.1997.

Lynn MacRitchie, „An all-women line-up for the Turner Prize": *The Sunday Times,* 18.06.1997, 23.

Tim Marlow, „Women on Top": *Elle Decoration,* 09/1997, 31.

John McEwen, „It has to be Parker": *The Sunday Telegraph,* 02.11.1997, 12.

Catherine Miller and Alasdair Palmer, „'I can't say what it means. That's your job'": *The Sunday Telegraph,* 22.06.1997, 23.

Steven Morris, „Bean bags, ear plugs and a video sitting still. Yes, it's Turner Prize", *Daily Mail,* 29.10.1997.

Jojo Moyes, „Hopefuls queue up for the ritual rubbishing", *The Independent,* 29.10.1997, 14.

„On the galopp": *Make 77,* Sept–Nov 1997, 4–5.

„One of these pieces is going to win £ 20,000. Which would you pick?": *The Journal (Newcastle),* 18.06.1997.

Joanna Pitman, „The fogeys bite back": *The Times,* 27.10.1997.

Nigel Reynolds, „A good idea, but is it really art?": *The Daily Telegraph,* 29.10.1997, 10.

Nigel Reynolds, „The Turner shortlist is for women only": *The Daily Telegraph,* 18.06.1997, 8.

Lina Saigol, „More than just a brush": *The Guardian,* 21.06.1997, 2.

Adrian Searle, „May the best woman win": *The Guardian,* 29.10.1997, 12–13.

Brian Sewell, „Turner's cruel history lesson", *Evening Standard,* 30.10.1997, 29.

Richard Shone, „Head Turners": *Artforum International,* 36, September, 73.

John Slyce, „The Odds on the Turner Prize": *Flash Art,* October, 49.

Birgit Sonna, „Der Rest ist Schwitzen": *Süddeutsche Zeitung,* 29.08.1997.

Robin Stringer, „The all-woman Turner Prize and not a painting in sight": *Evening Standard,* 17.06.1997, 7.

Helen Sumpter, „Eyes on the prize. Angela Bulloch deserves the Turner", *The Big Issue,* 02.11.1997, 33.

Christine Temin, „Human touch highlights machine art show", *Boston Globe,* 26.11.1996, E1, E7.

„The Turner Prize: Can we really call it art?": *The Week,* 28.06.1997.

Emily Tsingou, „Angela Bulloch": *Flash Art,* 193, March / April, 122.

Marina Vaizey, „Turner. Will the hype win again?": *Sculpture,* Autumn, 10–11.

„Variety is the spice of the Turner Prize exhibits", *The Herald (Glasgow),* 10.09.1997.

Sabine B. Vogel, „Setz dich, es geht los": *Frankfurter Allgemeine Zeitung,* 15.09.1997, 30

William Westwell, „The Turner Prize": *The News Line,* 23.09.1997, 9.

Andrew Wilson, „Band Art": *Art Monthly,* 210, October, 13–17.

Andrew Wilson, „Object Lesson": *Art Monthly,* 206, May, 1–7.

1996

Angela Bulloch, „On a Vanessa Beecroft Performance": *The Guardian,* 09.07.1996.

Andrew Cross, „Angela Bulloch": *Art Monthly,* 202, December / January, 25–26.

Mark Currah, „Angela Bulloch": *Time Out,* 20.11.1996, 66.

Pascaline Cuvelier, „Vital Signs": *Artforum International,* December.

Liam Gillick, „Une chaise qui devient un interrupteur / Like turning a chair into a switch": *Documents sur l'art,* 8, Spring, 27–31.

Peter Guth, „Menschenstürme im Licht-Atem": *Frankfurter Allgemeine Zeitung,* 20.03.1996.

Katharina Hegewisch, „Die Kunst der drei Grazien": *Wochenpost,* 13, 21.03.1996, 38–40.

Peter Herbstreuth, „In Leipzig": *Die Zeit,* 12.04.1996, 43–44.

Ralf Koch, „Schöne Begleiterin": *Frankfurter Rundschau,* 21.03.1996.

„Le bazar savant de l'art britannique": *Libération,* 10.10.1996, 30

Jade Lindgaard, „Angela Bulloch": *Les Inrockuptibles,* 11.12.1996, 27.

Andrew Mead, „Transatlantic artists telling tales of the riverbank": *The Architect's Journal,* 31.10.1996, 38–39.

Daniele Muscionico, „Meer aus Steckdose – Angela Bulloch in der Galerie Walcheturm": *Neue Zürcher Zeitung,* 18.09.1996, 54.

Michele Nicol, „Angela Bulloch": *Documents sur l'art,* 10, Winter, 24–25.

Hans Rudolf Reust, „Angela Bulloch – Szenen aus Betaville. Ein Gespräch": *Kunstforum International,* 135, 331–337.

Anuschka Roshani, „Künstlerwahnsinn": *Spiegel Extra,* 12, December, 6–12.

Manfred Sack, „Der Leipziger Kristallpalast“: *Die Zeit,* 12.04.1996, 43–44.

Nadia Schneider & Angela Bulloch, „Pillow Talk in Public Space / Öffentliches Kissengeflüster“: *Parkett,* 48, 154–161.

Claudia Spinelli, „Klick, klack, Meer an, Meer aus“: *Basler Zeitung,* 28.08.1996, 39.

Bianca Stigter, „Gedraag je!“: *NRC Handelsblad,* 08.03.1996.

Eric Troncy, „life / live“: *Art Press,* 219, December, 65–67.

Uljana Wuttig-Vogler, „Kunst zum Erfreuen und Streiten“: *Mitteldeutsche Zeitung,* 14.03.1996.

1995 Ester Adeyemi, „Weniger prägnante Positionen“: *Berner Volkszeitung,* 31.10.1995.

„Am Rande der Malerei II“: *Freiämter Tagblatt,* 08.11.1995.

„Am Rande der Malerei II“: *Swiss Agenda,* IX/1995.

Peter Anliker, „Altes Medium – aktuelle Ausprägung“: *Der Bund,* 27.10.1995, 9.

Mark Currah, „Angela Bulloch – Public Art Development Trust“: *Time Out,* October.

„Farbkleidungen und Farbmischungen“: *Berner Zeitung,* 27.10.1995.

Brigitte Oetker, „Kunst stiftet Messe-Identität“: *Die Welt (Sonderbeilage Leipzig),* 26.09.1995.

Rudolf Schmitz, „Wo bleibt der Körper?“: *Frankfurter Allgemeine Zeitung,* 23.11.1995.

1994 Michael Archer, „Angela Bulloch“: *Paletten,* 219, April, 5–7.

Patricia Brignone, „Angela Bulloch“: *Artefactum,* XI/52, Summer, 38–39.

Pascaline Cuvelier, „Les pièges iconoclastes de Bulloch“: *Libération,* 25.02.1994, 32.

Eva Karcher, „Die neunziger Jahre: Künstler als Forscher“: *Art – Das Kunstmagazin,* 7, 50–61.

Jeffrey Kastner, „Up against it“: *Art Monthly,* 178, July / August, 7–11.

Helena Papadopoulos, „Angela Bulloch, Sylvie Fleury“: *Flash Art,* XXVII/176, 103.

Éric Troncy, „Angela Bulloch. Centre de dréation contemporaine“: *Art Press,* 187, 85.

1993 David Lillingtion, „ON at Interim Art“ *Time Out,* 1169, 13.01.1993,

Eric Troncy, „Wall Drawings and Murals“: *Flash Art,* XXVI/170, May / June, 68–70

Giorgio Verzotti, „Angela Bulloch“: *Artforum International,* April, 105.

Giorgio Verzotti, „Aperto 93“: *Flash Art,* XXVI/172, October.

Yvonne Volkart, „Eleven Women“: *Flash Art,* XXVI/172, October.

Mark White, „ON at interim Art Limited“: *Twister,* 1, 9.

1992 Francesco Bonami, „Angela Bulloch – Lonesome Comedy“: *Flash Art,* XXV/163, March / April, 96–97.

Francesco Bonami, „Plastic Fantastic Lover, Master and Mistresses“: *Flash Art,* XXV/162, January / February, 125.

Nicolas Bourriaud, „The Work of Art in the Age of Ecological Recycling“: *Flash Art,* XXV/167, November / December, 60–63.

Dan Cameron, „Don't look Now“: *Frieze,* 3, 4–8.

Sonia Criton, „il faut construire l'hacienda“: *Art Press,* 167, March.

Catherine Grout, „Il faut construire l'hazienda“: *Arte Factum,* April / May, 36.

Tom Kalin, „Plastic Fantastic Lover“: *Artforum International,* January, 120.

Elisabeth Lebovici, „L'art piégé à l'anglaise“: *Libération,* 07.07.1992, 33.

Gerhard Mack, „Mehrerlei Comics“: *Kunst Bulletin,* November, 20–23.

„Plastic Fantastic Lover“: *Arts Magazine,* January, 81.

Noemi Smolik, „Müll, Kunst und Leben“: *Kölner Stadtanzeiger,* 17.07.1992.

Jean-Pierre Thibauda, „Mais qu'en pense le kangorou?“: *Libération,* 26.02.1992.

Eric Troncy, „London Calling“: *Flash Art,* XXV/165, Summer, 86–89.

Benjamin Weil, „Angela Bulloch“: *Flash Art,* XXV/166, October.

Benjamin Weil, „Cut it out“, *Frieze,* 4, April / May, 34–40.

1991 Dalya Alberge, „Making their mark“: *The Independent,* 17.12.1991, 15.

„Farbe per Funksignal“: *Kölner Illustrierte,* January.

Peter Frank, „Common Market, Boyd Webb, Dominique Blain, Anthony Gormley“: *LA Weekly,* 16.02.1990, 120.

Liam Gillick, „The Placebo Effect – Some Art in Britain“: *Arts Magazine,* May, 56–59.

James Hall, „Pure Gold(smiths)“: *New Statesman & Society,* 02.08.1991, 50–51.

Jean-Yves Jouannais, „no man's time“: *Art Press,* October, 95.

Sarah Kent, „Breaking Ground“: *Time Out, 1095,* 14.08.1991.

Jose Lebrero Stals, „Angela Bulloch“: *Flash Art,* 157, March / April, 147–148.

William Packer, „The diminishing value of novelty“: *Financial Times,* 06.08.1991, 21.

Roberta Smith, „Plastic Fantastic Lover“: *The New York Times,* 01.11.1991.

Eric Troncy, „Soft Touch“: *Artscribe,* September, 54–57.

Eric Troncy, „No Man's Time“: *Flash Art,* 161, November / December. 119–120.

Giorgio Verzotti, „No Man's Time“: *Artforum International,* November, 148–149.

Oliver Zahm, „Rêve, Fantasie“: *Artforum International,* May.

1990 Marjorie Allthorpe-Guyton, „Seven Obsessions“: *Artforum International,* November.

Mark Currah, „Angela Bulloch“: *City Limits,* 22.03.1990, 87.

William Feaver, „Exorcise classes for the obsessed“: *The Observer,* 02.09.1990.

Liam Gillick, „Seven Obsessions“: *Flash Art,* 155, November / December.

Andrew Graham-Dixon, „The World of Interiors“: *The Independent,* 28.08.1990.

James Hall, „Portrait of the artist as a dedicated fruitcake“: *Sunday Correspondent,* 26.08.1990.

Rose Jennings, „Angela Bulloch. Interim Art“: *City Limits,* 08.03.1990.

Jutta Koether, „Three from England“: *Flash Art,* January / February, 138.

Jutta Koether, „It's a Maggot Farm, the B-Boys and Fly Girls of British Art: Five Statements and a Conversation“: *Artscribe,* 84, Summer, 56–62.

„Letter from London“: *Contemporanea,* March.

David Lillington, „Angela Bulloch. Interim Art“: *Time Out,* 21.03.1990.

Martin McGeown, „Angela Bulloch. Interim Art Gallery“: *Art Press,* 148, June.

Andrew Renton, „Angela Bulloch: Artist“: *Blitz,* 87, March, 6.

Andrew Renton, „Plotting“: *Performance,* 62, November, 36–45.

Andrew Renton, „Angela Bulloch“: *Flash Art,* 153, Summer, 152.

Noemi Smolik, „Werk ohne Schöpfer“: *Kölner Stadtanzeiger,* 12.12.1990

„Werk ohne Schöpfer. Angela Bulloch in der Galerie Esther Schipper“: *Kölner Stadtanzeiger,* 07.12.1990.

Olivier Zahm, „Angela Bulloch. Galerie Claire Burrus“: *Art Press,* 150, September.

1989 Angela Bulloch, „40 Under 40: The New Generation in Britain“: *Art & Design,* V/6–7.

Angela Bulloch, „FREEZE“: *Art & Design,* V/3–4, 52–53.

Impressum
Imprint

Diese Publikation erscheint anlässlich der Ausstellung
This catalogue is published in conjunction with the exhibition

The Space That Time Forgot

16. Februar – 18. Mai 2008
February 16 – May 18, 2008

Städtische Galerie im Lenbachhaus und Kunstbau, München
Luisenstrasse 33
80333 München
Deutschland / Germany
Tel. +49 89 233 320-00
Fax +49 89 233 320-03
www.lenbachhaus.de

Ausstellung / Exhibition Angela Bulloch, Matthias Mühling
Assistenz / Assistant Felix Prinz
Organisation / Coordination Susanne Böller, Karin Dotzer
Kommunikation / Communication Claudia Weber, Jonna Gaertner
Technische Leitung / Technical administration Andreas Hofstett

Katalog / Catalogue
Herausgeber / Editor Matthias Mühling
Konzept / Concept Angela Bulloch
Lektorat / Copyediting Susanne Böller, Axel Lapp, Matthias Mühling, Felix Prinz
Biografie, Bibliografie / Biography, Bibliography Axel Lapp
Übersetzungen / Translations Greg Bond, Tradukas (Vorwort), James Gussen (Diederichsen), Michael Eldred (Mühling)
Fotos / Photos Carsten Eisfeld, Simone Gänsheimer (5)
Gestaltung / Design Carsten Eisfeld
Lithographie / Lithography Eberle & Eisfeld | Berlin
Druck / Printing DZA Druckerei zu Altenburg GmbH

Erschienen im / Published by
Verlag der Buchhandlung Walther König, Köln
Ehrenstraße 4, 50672 Köln
Tel +49 (0) 221 / 20 59 6-53
verlag@buchhandlung-walther-koenig.de

Die Deutsche Bibliothek – CIP-Einheitsaufnahme
Ein Titelsatz für diese Publikation ist bei Der Deutschen Bibliothek erhältlich

Printed in Germany

Distribution:

Schweiz / Switzerland
AVA, Verlagsauslieferungen AG
Centralweg 16, Postfach 27,
CH-8910 Affoltern a. A.
Tel +41 (0) 1 762 42 00
Fax +41 (0) 1 762 42 10
a.koll@ava.ch

UK & Eire
Cornerhouse Publications
70 Oxford Street
GB-Manchester M1 5NH
Tel +44 (0) 161 200 15 03
Fax +44 (0) 161 200 15 04
publications@cornerhouse.org

Außerhalb Europas / Outside Europe
D.A.P. Distributed Art Publishers, Inc., NY
155 Sixth Avenue
New York, NY 10013
Tel +1-212-627-1999
Fax +1-212-627-9484

ISBN 978-3-86560-426-2

Dank
Thanks

Die Künstlerin dankt allen, die bei dieser Ausstellung und bei diesem Katalog, sowie zu Angela Bullochs Arbeiten und Projekten hilfreich zur Seite standen.

The artist would like to thank all those who have helped her with this exhibition and the catalogue, Angela Bulloch's works and projects.

Helga de Alvear, Simon Baier, Isabel Bernheimer, Susanne Böller, Vittorio Dapelo, Diedrich Diederichsen, Carsten Eisfeld, Antonio Ferrara, Helmut Friedel, Holger Friese, Nunzia & Vittorio Gaddi, Susanne Gaensheimer, David Grubbs, Cristina Guerra, Christoph Gurk, Andreas Hofstett, Thomas Huesmann, Michael Iber, Axel Lapp, Barbara-Brigitte Mak, Sigrid Melchior, Matthias Mühling, Max Müller-Heidelberg, Eva Presenhuber, Felix Prinz, Attila Saygel, Sonja Schamberger, Esther Schipper, Lorenz Schreiber, Rainald Schumacher, Michael Stadler, Micheline Szwajcer-Tob, Claudia Weber, Christophe Wiesner.